Light

Light

RICHARD MORRIS

The Bobbs-Merrill Company, Inc.
Indianapolis   New York

Copyright © 1979 by Richard Morris

All rights reserved, including the right of reproduction
in whole or in part in any form
Published by The Bobbs-Merrill Company, Inc.
Indianapolis   New York

Designed by Rita Muncie
Manufactured in the United States of America

Second Printing

Library of Congress Cataloging in Publication Data

Morris, Richard, date.
  Light.

  Bibliography: p.
  Includes index.
  1. Light.   I. Title.
QC358.5.M67   535   78-11206
ISBN 0-672-52557-7

Contents

List of Illustrations

One:
Light: The Search for Understanding

During the fifth century B.C., the Greek philosopher Empedocles tried to explain sight by proposing that light was something which emanated from the eye and struck objects in the field of vision. The interaction between the light and the objects was somehow transmitted back to the viewer's mind. The result was vision.

Of course Empedocles was wrong; the Greek philosophers were often wrong. We admire them, not because they found the solutions to the majority of the problems they posed, but because they realized that the problems existed. If they failed to find all of the answers, they did succeed in inventing the questions. And that was an achievement.

Strangely enough, Empedocles' question, "What is light?" proved to be one of the most difficult of all queries. More than two thousand years were to pass before it could be answered with any degree of certainty. In one age after another, humanity's best minds struggled with this problem. Whenever it seemed that a solution had been found, someone was sure to demonstrate, within a very short time, that the answer was the wrong one. Empedocles' question, which had seemed so simple, turned out to be more baffling than many abstruse philosophical questions. For example, by the thirteenth century Thomas Aquinas thought he had solved some

1

of the problems posed by Plato's universals. Later the Church canonized him for his work. But no one either in Aquinas's time or in the centuries immediately following had any idea of the true nature of light.

It wasn't until the twentieth century that scientists succeeded in gaining a fairly adequate understanding of light, and then they had to create a revolution in physics in order to do so. The fundamental theories on which all modern physics is based—relativity and quantum mechanics—grew directly out of attempts to understand the nature of light.

The twentieth-century upheaval in physics is not the only revolution that has been associated with a preoccupation with light. For example, the assertion that the impressionist movement in painting grew out of attempts to interpret the effects of light in the open air on canvas has become almost a cliché. The painters of the Renaissance were equally concerned with light. The Renaissance invention of chiaroscuro—the modeling of forms by the use of light and shade—had an impact on art that was at least as revolutionary as any of the innovations of Monet or Renoir or Pissarro.

For thousands of years light has been a widely used symbol in religion and mythology. The writers of the Gospels were so impressed by the revolutionary nature of the teachings of Jesus that they called him "the Light of the World." In this they unknowingly echoed the ancient Persian prophet Zarathustra, the founder of one of the world's most influential religions.

The doctrines of Zarathustra have few adherents today except among the Parsees, a small sect in India. Everyone, however, is familiar with the mythic symbolism that Zarathustra created: the identification of light with good and darkness with evil. This symbolism has pervaded Western culture for the last two thousand years. Not only is it reflected in such terms as "Christ the Light" and "Prince of Darkness"; it is even seen in old western movies, where the hero always rides a white horse and the villains are dressed in black. Such associations are so much a part of our thinking that we would find it very difficult not to associate white with purity and black with evil, the supernatural and ignorance.

There are entire technologies associated with light. Somehow, things involving light always seem to be more astonishing than feats performed with electricity or sound or even atomic energy. Radio, the phonograph and the telephone were all invented during the nineteenth century. And yet it was the invention of photography that caused the greatest popular sensation. The thought that light could be used to reproduce an image on a photographic plate seemed so astonishing that to some the very idea was blasphemous.

Today we are somewhat more sophisticated; we no longer feel particularly awed by new technical marvels. Computer technology, nuclear energy and even space flight have come to seem commonplace. If we are astonished by any feat, it is what can be accomplished with lasers and holograms. Anything that is done with light seems to have something magical about it.

A very long time ago—no one really knows how long—the Chinese devised the yin-yang symbol to express the fact that opposites always seem to be intertwined with each other. One cannot speak of light without considering darkness. Similarly, *hot* implies the existence of something called *cold*; wisdom cannot exist without its opposite, ignorance; and if there were no good, we would never be able to conceive of evil.

Scientists have had to deal with a pair of yin-yang opposites in their attempts to understand light. An observation of darkness established the wave theory during the nineteenth century; when it was shown that light could interfere *destructively*, by creating dark spots or bands, the opponents of the wave theory had to admit they were wrong. Today we encounter a similar situation: studies of light have caused scientists to hypothesize those strange astronomical objects known as black holes.

Black holes are fascinating because they are completely dark. They have gravitational fields so strong that no light is permitted to escape them; it is scientifically accurate to describe them as "black holes in space." Since they suck up everything that comes near them, light as well as matter, it is impossible to see them; their existence can be established only by inference. Yet they are phenomena of great scientific importance, and they have elicited a great deal of theoretical speculation, some of it quite fantastic.

Light and darkness also enter into theories about the origin and fate of the universe. It has been suggested, for example, that the universe may eventually condense into a massive black hole. Although that is no more than speculation, we can be reasonably sure that the universe began with an enormous explosion in which all of space was permeated by light.

This is the so-called big bang theory of the creation of the universe which by now is almost universally accepted. The theory received sensational confirmation in 1964, when it was discovered that the light from the primordial fireball could still be detected as it approached the earth from all directions of space. Although twenty billion years had passed since the creation, the light was still there.

Light, then, is intrinsically a subject of great interest. But it is more than that, I think. In a way, the story of light provides a kind of paradigm of mankind's entire intellectual and cultural history. The tale is so dramatic and so varied and it touches on so many different subjects that an examination of the ways in which men and women have looked at light can give us a picture of humanity's millennia-long struggle toward understanding itself and the universe.

The story begins long before the dawn of recorded history, perhaps even before there was any such animal as man, but when intelligence had evolved sufficiently so that some creature or another became aware of the life-giving light and heat that flowed to the earth from the sun. However, since events in so distant a past can be dealt with only in speculation, the story presented in this book will begin with the Persian prophet Zarathustra, who lived approximately three thousand years ago.

Zarathustra was apparently the first to perceive a cosmic symbolism in the two primal opposites, light and darkness. This symbolism was so compelling that, in time, it became part of the teachings of Judaism and Christianity. It was reflected in the doctrines of the sects which produced the Dead Sea Scrolls and in the New Testament Gospels. It then became part of the Gnostic and Manichaean heresies, and eventually it was suppressed. The symbolism persisted in literature, however, if not in doctrine. Dante makes numerous references to it in his *Divine Comedy*, and Milton places great emphasis on both light and darkness in *Paradise Lost*. Other authors, although perhaps not so preoccupied with light and darkness as Dante and Milton, frequently refer to the two opposites. One of the best known poems in the English language is "The World" by Henry Vaughan, which begins, "I saw eternity the other night,/Like a great ring of pure and endless light." And in Shakespeare's *Othello*, the protagonist, brooding about his contemplated murder of Desdemona, says, "Put out the light, and then put out the light." As the soliloquy continues, Othello finds himself continually drawn back to the word *light* and to the idea of extinguishing it.

Within a relatively short time after Zarathustra's death, the Greeks became interested in light. They, however, were not really concerned about symbolism. The Greeks endeavored to gain an understanding of the universe by using their powers of reasoning. Euclid, Plato and Aristotle, and the pre-Socratic philosophers as well, pondered the question of the nature of light. Sometimes they confused light as a physical entity with the subjective quality, sight, but they cannot be blamed for that: to the Greeks, "optics" was the science of vision.

The Greeks reached conclusions which we realize today were not very accurate. However, they did pass on their interest in light to Arab scholars, who preserved classical knowledge through that period which is known, at least in the West, as the Dark Ages.

At the beginning of the Renaissance, a renewed interest in the subject of light was prompted by the desire of painters to represent light and perspective more accurately. Leonardo's notebooks, for example, contain a number of passages on the use of light in painting. And before long Galileo was attempting to find a way to measure light's velocity.

Techniques of representing the effects of light on canvas quickly became part of the body of knowledge available to any reasonably competent painter. No one, however, had been able to figure out exactly what light *was*. When the problem was finally attacked, it proved to be an extraordinarily difficult one. The French philosopher Descartes studied light and correctly gave the laws of reflection and refraction, but his ideas about the composition of light were soon shown to be wrong. The English scientist Isaac Newton proposed another theory of light. Although Newton did succeed in demonstrating that white light is a composite of the various colors of the spectrum, his ideas about the nature of light were incorrect. Newton's authority, however, was so great that his contemporaries accepted his theory of light without reservations, thus creating an orthodoxy that was to prove inimical to progress toward a correct theory.

In 1801 a young English scientist named Thomas Young gave a lecture before the Royal Society in which he described an experiment he had conducted which seemed to show that light was made up of waves. Young's theory was greeted with derision because he had contradicted the great Newton.

Obviously, Young's contemporaries thought, his ideas must be absurd.

Newton had said that light was composed of streams of particles. Although this idea did eventually lose out to Young's wave theory, the process took decades. Finally, two Frenchmen vindicated Young's theory, but in the land of Newton no one seemed willing even to perform the necessary experiments.

The wave theory had hardly become established when an earth-shaking event took place: Louis J. M. Daguerre announced the invention of the first workable photographic process, which became known as the daguerreotype. Within a few years, photography had spread throughout the world.

Before the nineteenth century was much more than half over, a group of young Parisian painters who were preoccupied with representing the effects of light began to challenge the authority and standards of the French Academy. The critics and the public referred to them derisively as impressionists. The impressionists accepted the appellation as a badge of honor and went on to create the revolution that made modern art possible.

Meanwhile, discoveries were being made in physics at an increasingly rapid pace. Eventually, James Clerk Maxwell was able to show that light was a kind of electromagnetic radiation. A German scientist confirmed the theory by producing another kind of electromagnetic wave: radio. Finally, Marconi applied the principles of the discovery to produce wireless telegraphy.

By the time the nineteenth century reached an end, scientists were sure they had discovered all of the fundamental laws of nature. There was nothing more to be done, they thought, except find the solutions to a few annoying little puzzles. One of these had to do with the emission of light. A scientific revolution was to take place before an answer could be found to that puzzle.

By 1905 this revolution had begun to gain momentum, and decades were to pass before the pace would slacken. Einstein made sure of this by announcing, in that same year, his theory of relativity and his discoveries about the nature of light.

Light, it seemed, was made up of neither waves nor particles alone; it somehow managed to be both simultaneously. However, contradictory this seemed, it had to be accepted; the experimental evidence would not allow any other interpretation.

Soon it was discovered that matter too had a dual nature; it also appeared to be made up sometimes of waves and sometimes of particles. One discovery followed another until in the 1970s the world of the physicist was populated with such strange objects as the quark and the quasar, the big bang and the black hole. Strangely, many of these new entities seemed to be bound up, in one way or another, with the idea of light.

Some scientists believe that we will never discover the ultimate nature of physical reality, if indeed there is any such thing. They liken the job of doing physics to the peeling of an onion. As each successive layer is removed, as we conduct new experiments and formulate new theories, we penetrate more deeply into the nature of reality. But they say there is no guarantee that this process will ever come to an end. It might be that, however many layers we peel, still more will remain. We may invent more sophisticated theories and look more deeply into the atom and its constituent particles without ever reaching any ultimate truth. As the puzzles of physics are solved, we may find that we have only replaced them with new ones.

Perhaps writing a book is also like peeling an onion. It is no more possible to know or tell everything about a subject than it is to know everything about the universe. We'll never understand everything about the manner in which the mythical symbolism of light versus darkness came about. It would be as absurd to think we could know everything that was going on in Claude Monet's mind as he tried to recreate the effects of light on canvas as it would be to think that physics will ever pronounce the last word on the nature of quarks or photons of light or black holes.

A few books, however, have succeeded in pulling back the first few layers of the onion. The author has attempted to do something like that with the subject of light in the chapters that follow.

Two:
Lords of Light and the Prince of Darkness

For more than two thousand years it has seemed natural, at least to those of us who have lived within the Judaeo-Christian tradition, to equate God with light and evil with darkness. For example, the Gospels say that Jesus is the Light of the World. St. Paul's conversion begins when he is blinded by a great light from heaven. In Christian art the heads of Jesus and the various saints are often surrounded by halos. And literary allusions to divine light, such as those in Dante's *Paradiso*, are too numerous to mention. The identification of God with light is so pervasive that we don't even think it odd that the story of the creation as given in Genesis should allow three days to pass between the time when God says "Let there be light" and His creation of the sun and moon.

The Devil, on the other hand, is the Prince of Darkness. He is described in folklore as a dark or black man; hell, the kingdom over which he presides, is, in spite of its fires, a dark and dismal place. In the Gospels, sinners are said to "walk in darkness." We habitually think of evil as something dark, and speak of black hearts, black magic and dark thoughts or deeds. The identification of evil with the color black is common even in black Africa, although in Mozambique there is a demon called Muzungu Maya, "wicked white man."

11

But is the identification of God (or good) with light and evil with darkness as natural as it seems? It is missing from many ancient mythologies. For example, the Egyptian god Osiris, who is murdered by his evil brother Set, is not equated with light, nor is his killer with darkness; the color associated with Set is not black but red. The Babylonian god Marduk is not identified with light, nor is the Greek Zeus or the Canaanite god Baal. It is true that there have been numerous sun gods, but may it not be the warmth of the sun rather than its light that is seen as the giver of life? Every now and then we do encounter a god such as Quetzalcoatl who is thought of as being very fair-skinned, but it is equally easy to find gods like Krishna who are described as being dark.

The dichotomy of light and darkness has probably occurred to many people in many ages. But it was the ancient Persians in the second or first millennium B.C. who elevated it into a cosmic myth. In the religion preached by the prophet Zarathustra (or Zoroaster, as the Greeks called him), light and darkness become cosmic principles which vie with each other for supremacy in a battle that will continue until the end of the world.

Of all the founders of the world's great religious systems, Zarathustra is perhaps the most enigmatic. We know that he was a reformer of an ancient faith that was probably similar to the Vedic religion of India, but we have no direct knowledge of what that faith was. We have written records of Zarathustra's teachings; his utterances are recorded in the "Gathas," which are the earliest written documents of Persian culture. However, we can't always be sure of the meaning of this prophet's words. Because the "Gathas" are written in a difficult language that has not been completely deciphered, the scholars who study them often disagree as to what they say. To make things worse, three-quarters of the *Avesta*, the great book of Zoroastrianism of which the "Gathas" are a part, is lost. And finally, we can't even be sure when Zarathustra lived. Although the date traditionally assigned to his birth, 630 B.C., seems a fairly reasonable one, some experts date him as early as 1500, others as late as 550 B.C.

Fortunately, we possess writings of a later date than the "Gathas" which give clear accounts of Zoroastrian doctrine. These tell us that Zoroastrianism was a dualistic religion which viewed the cosmos as a battleground on which Ahura Mazdah, the Lord of Light, fought Angra Mainyu, the Spirit of Darkness. Furthermore, we have evidence, especially since the discovery of the Dead Sea Scrolls, of the influence of Zoroastrianism on Judaism and Christianity. It seems to be fairly well established that Zarathustra is the person responsible for the symbolism of light and darkness that still influences us so greatly.

Before we examine the ways in which Zoroastrian doctrines influenced the Judaeo-Christian tradition, it would be a good idea to take a closer look at those doctrines. In the following discussion I will refer to Ahura Mazdah as God and Angra Mainyu as the Devil. They are close enough to our God and Devil that no paradox will be introduced by simplifying the terminology.

According to Zoroastrianism, in the beginning God created a spiritual world that was inhabited by spiritual beings. The Devil, who was apparently

Relief of Ahura Mazdah, the ancient Persian god of light.
Courtesy Oriental Institute, University of Chicago.

rather dull-witted, dwelt in his own kingdom of Endless Darkness for three thousand years before he even became aware of God's realm of Endless Light. But when he did notice it, he rushed up at once to destroy it. When God made an offer of peace, the Devil turned it down, believing that God had proposed a truce because He knew Himself to be helpless.

God in His omniscience had foreseen this. He answered the Devil by proposing that the two of them do battle for nine thousand years, knowing that at the end of that time the Devil and all his demons would be destroyed. He foresaw that in the last three thousand years of time, beginning with the birth of Zarathustra, the Devil would be rendered powerless.

God then hurled the Devil back into darkness, where the latter lay unconscious for three thousand years. While the Devil was asleep, God undertook a second creation, the earth. After making sky, water, the solid earth, plants and cattle, He created Gayomart, the First Man, and gave him the Lone-created Bull as a companion. Finally, God created fire, its brilliance derived from His own kingdom of light.

When the Devil awoke, he went forth with all his demons to attack the earth. He tore open the sky and created the moving planets in order to destroy the fixed order of the constellations. He polluted the waters, caused drought, and introduced biting and poisonous reptiles, frogs and lizards. He caused a blight to fall on the plants that God had created, and bored a hole into the center of the world (which was later to become hell). Finally, he mixed darkness and smoke with God's fire so that the entire earth became dark. He killed the Lone-created Bull and set upon the First Man, slaying him with the help of a thousand demons.

Again, all of this had been foreseen by God, who closed up the gap in the sky and trapped the Devil within His second creation. God and His angels then descended to do battle. At the end of ninety days, they cast the Devil down into hell.

The Devil, although he had been hurled through the hole that leads to the center of the world, considered this the hour of his greatest triumph because he had successfully defiled God's creation. He had not yet realized that he had unwittingly allowed himself to be caught in a trap from which there would be no escape.

Now, when the First Man had died, eight kinds of precious metal had flowed out of his body: gold, silver, iron, brass, tin, lead, quicksilver and adamant. The gold remained in the earth for forty years. At the end of that time, it emerged in the form of a single rhubarb stalk. When the stalk was grown, it split into two, becoming Mashye and Mashyane, the Zoroastrian equivalents of Adam and Eve.

By this time some of the damage inflicted on the world had been repaired. God had not, however, been able to restore it completely; as long as the Devil remained within His creation, it would be defiled. God instructed Mashye and Mashyane to obey His commandments and not to worship demons. If they disobeyed, the demons would gain power and humanity would come to grief.

Perhaps the Zoroastrian God expected to be obeyed. But as it turned out, he had no more luck than the Hebrew God when the latter gave admonitions to Adam and Eve. Mashye and Mashyane immediately fell into temptation, not once but several times. At almost their first opportunity, they declared the Devil to be the creator of water, the earth and the plants. They engaged in forbidden sacrifices and committed other sinful acts. As a consequence, the power of the demons became so great that they were able to make the human couple sexually impotent for fifty years. When Mashye and Mashyane finally overcame this problem and produced a pair of twins, however, they gave no indication of having reformed; they ate the children immediately.

Patiently, God took away the "sweetness of children" so they wouldn't devour future offspring. God's ploy turned out to be successful. The incompetent Mashye and Mashyane (who weren't really bad, just stupid) bore an additional seven pairs of twins, who proceeded to people the earth.

According to Zoroastrian doctrine, every man and woman plays a part in the struggle between God and the Devil. Sinful acts cause the demons to become more powerful, while moral behavior contributes to the ultimate victory of Light over Darkness. Every human being is faced with a choice; he or she is either on the side of good or is a partisan of evil.

At the end of time, when the nine-thousand-year war is nearing an end, the Devil will become weak enough that he can be defeated in an apocalyptic battle. Just before this happens, however, the forces of Darkness will seem to be gaining in power as they prepare for the concluding struggle (this formula, incidentally, is still followed in western movies, where the bad guys always appear to be winning just as they are on the point of being defeated).

After the Devil is overthrown, the messiah Soshyans will come to oversee the resurrection of the dead, and all humanity will enjoy eternal life in God's kingdom of Endless Light. Even those souls that have been languishing in hell will be saved after they undergo a three-day purgation in molten metal. In Zoroastrianism, there is eternal life but no eternal damnation.

Unlike Buddhism and Christianity, Zoroastrianism never attempted to become a world religion. In Iran it was eventually supplanted by Islam, and today it survives only among the Parsees of India, a sect that numbers only about a hundred thousand. Nevertheless, the influence of Zoroastrian ideas has been immense, especially in Christianity. The Christian Devil is thought to have his origins in Zoroastrianism, as does the thesis found here and there throughout the New Testament that the forces of Darkness, under the leadership of Satan, are warring with Light. Even the halo (which is found in Buddhist and Islamic art as well as Chris-

The halo, which is found in Christian, Buddhist and Islamic art, was originally a Zoroastrian motif called the *xvarenah*, symbolizing the heavenly fire, a luminous life-force which is communicated to man. The painting pictured below is a fresco. *Courtesy Gabinetto Fotografico Del Museo Civico Di Padova.*

tian) is a Zoroastrian motif. Called the *xvarenah*, it symbolizes the heavenly fire, a luminous life force that is communicated to man.

Admittedly, the existence of a Christian Devil and a Zoroastrian Devil doesn't necessarily imply that one must be identified with the other. If the Gospel and Epistles of St. John and, to a lesser extent, the other books of the New Testament make numerous references to divine light, this doesn't necessarily mean that the New Testament writers were influenced by Zoroastrian ideas. It is necessary to have confirming evidence before we can reasonably come to such conclusions.

Many scholars believe that this evidence exists. The Dead Sea Scrolls, for example, show that Zoroastrian ideas were widespread during the centuries immediately preceding the beginning of the Christian era. For example, one of the scrolls, the *Manual of Discipline*, states a doctrine that is very similar to the tenets of Zoroastrianism. The *Manual* speaks of two spirits, one good and the other evil, which have their respective origins in the "Fountain of Light" and the "Wellspring of Darkness." The two spirits are not quite identical with the Zoroastrian God and Devil, Ahura Mazdah and Angra Mainyu; they are described as having been created by God and as being subject to Him. However, like the Zoroastrian God and Devil, they do war with each other for the world and for individual human souls in a prolonged cosmic battle. Victory will eventually be given to the Spirit of Light, but only after a bitter struggle.

Another scroll prophesies an apocalyptic war between the "Sons of Light" and the "Sons of Darkness." The final battle is described in excruciating detail. Not only does the scroll describe the formations in which the troops will be drawn up; it also gives the inscriptions that will be found on the standards of the various regiments, brigades, companies, platoons and squads. Finally, it quotes a lengthy speech that must be made by the high priest before the fighting can start. When it does begin, the Sons of Light are of course victorious. After all, the Sons of Darkness have a leader who, unlike the high priest, is not known for his ability to preach rousing sermons.

It is easy to find parallels between the Dead Sea Scrolls and the New Testament Gospels. For example, in the twenty-fourth chapter of the Gospel according to St. Matthew, Jesus speaks of a coming age of conflict and destruction. "Nation shall rise against nation," He says, and prophesies that there will be "false Christs and false prophets" and finally an "abomination of desolation."

To many scholars this suggests that Jesus believed that the battle between the Spirit of Light and the powers of Darkness was nearing a climax, that Evil would soon make its final bid for supremacy. This would lead to a terrible conflict; in every man and woman the forces of Darkness would increase their struggle against Light and Truth.

Although it is true that the New Testament and early Christianity are permeated with the idea that the end of the world is near and that a terrible struggle is imminent, it is difficult to say what Jesus believed with any degree of certainty. Most Biblical scholars believe that the Gospels were written approximately forty years after his death by men who had not known Him. The writers who had to draw on an oral tradition that had grown up in the years

following the Crucifixion didn't know Jesus'
thoughts. However, it would not be reasonable to
think that he could have been completely oblivious
to the apocalyptic ideas that were so much in the air
at the time.

Zoroastrian ideas had their influence on Judaism
too, but to a lesser extent than on Christianity.
Nevertheless, I think it is easy to show that the
Jewish and Christian conception of the Devil as
God's powerful opponent derives from Zoroas-
trianism, that Satan is none other than Angra
Mainyu.

It is sometimes necessary, however, to separate
Jewish ideas from Christian. If we allowed ourselves
to confuse the two, it would be impossible to trace
the development of the idea of the Devil with any
accuracy. In discussing Judaism, we must not identify the Devil with the serpent in the Garden of
Eden. This is a Christian interpretation, one that is
bound up with the doctrine of original sin. The
conception of Satan as the personified power of evil
is a relatively late development in Judaism; there is
no Devil named in Genesis at all unless we read him
into it.

The serpent is not described as being a Devil, but
only as a creature that is "more subtil than any beast
of the field." The impression is given that it is the
serpent who bears the ultimate responsibility for the
Fall, but this is part of no cosmic plot.

If we don't find any Devil in the earliest written
parts of the Old Testament, that is only natural.
The Jews had been worshipping their god Yahweh
(misspelled as "Jehovah" in the King James Version)
for a very long time before they came into contact
with the Persians and Zoroastrianism. As a result,

even though Palestine was part of the Persian empire for two centuries, neither the identification of God with light nor the emphasis on the power of the Devil ever became as strong in orthodox Judaism as it later did in Christian thought.

Many of the Old Testament references to Satan do not refer to the Devil at all. *Satan* is a Hebrew word meaning "adversary." In the book of Job, for example, Satan is included with the "sons of God," members of the heavenly court. He does not seem to be an evil spirit, only a kind of public prosecutor or (if the reader will permit the pun) Devil's advocate. This is established at the beginning of the book: in Job 1:6 we read, "Now there was a day when the sons of God came to present themselves before the Lord, and Satan came also among them."

Satan, it turns out, is a rather cynical observer of human behavior. When Yahweh makes mention of Job's piety, Satan objects on the grounds that anyone would be pious if he had received as many material blessings as Job. In the subsequent testing of Job, Satan acts only with Yahweh's permission; he does not act on his own. He is not yet Satan the Prince of Darkness, but only "a satan," an adversary. He is nothing like Angra Mainyu, who consciously strives to defile God's creation.

There are other instances of the use of the word *satan* in the Old Testament where similar meanings are obviously intended. For example, the Philistines reject David as an ally because they are afraid he might switch sides and become their satan, or adversary (I Samuel 29:4), and Solomon exults that the Lord has given him "neither adversary [satan] nor evil occurrent" (I Kings 5:4).

In the later Old Testament books, however, Satan

becomes the Devil. That this transformation shows the influence of Zoroastrianism finds confirmation in the fact that the Old Testament contains some seemingly direct references to Zoroastrian doctrine. In order to put these in an understandable context, however, it will be necessary first to say a few words about Cyrus the Great.

When the Persian king Cyrus the Great overthrew the Chaldean empire of Babylonia in 538 B.C., he released the Jews from captivity and resettled them in Palestine. Shortly thereafter he removed the captured images of the deities of numerous Near East cities from the temple in Babylon and restored them to their original sites. At about the same time he commanded that the temple in Jerusalem be rebuilt.

Since Cyrus was a Zoroastrian, we can reasonably suppose that he considered himself to be acting under the guidance of Ahura Mazdah. Nevertheless, it was to Yahweh that the Jewish priests gave all the credit. The unknown prophet who wrote the last twenty-six chapters of the book of Isaiah (the so-called Second Isaiah) echoes this sentiment when he calls Cyrus "the anointed of Yahweh" and tells us that the Lord has said of the Persian king, "He is my shepherd." Then, as if all this were not enough the prophet quotes Yahweh as saying, in opposition to Zoroastrian doctrine, "I form the light and create darkness: I make peace and create evil: I the Lord do all these things" (Isaiah 45:7).

Nevertheless, Zoroastrian ideas had an influence. During the post-Exilic period, the idea of Satan as God's powerful adversary gradually became established in Jewish thought. The Jews of earlier times, though they had possessed an elaborate demonol-

ogy, had apparently not felt the need for a personified power of evil. Those whom Cyrus had released
from captivity, on the other hand, seem to have
turned Angra Mainyu into God's supernatural adversary Satan.

The first Old Testament mention of Satan as
God's evil opponent can be found in the twenty-first
chapter of I Chronicles. We read that Satan tempts
David into conducting a census, a sin which
Yahweh requites by sending a plague. This is exactly
the same story that is told in II Samuel 24:1–15. The
only difference is that in Samuel it is Yahweh himself who tempts David into numbering his people.
Apparently the writer of the earlier book felt no
qualms about ascribing temptation to Yahweh. Or
perhaps he thought he had no one else to blame it
on. This proved to be unacceptable to the post-
Exilic author of Chronicles, who, perhaps acting
under the influence of Zoroastrian ideas, ascribed
the responsibility for David's act to Satan.

The tradition which identified all evil with Satan
was firmly established during the Hellenistic age (a
period approximately two hundred years in length,
dated from the death of Alexander the Great in 323
B.C.). The apocalyptic writings of the age contain a
dualistic demonology that certainly seems to derive
from Zoroastrian ideas. The leader of the powers of
Darkness is called variously Satan, Mastema,
Semyaza, Azazel, and Beliar or Belial. The ideas of
Satan as a fallen angel and of a primordial war in
heaven are developed. These appear again in the
New Testament book of Revelations, where they
appear to be combined with an ancient myth concerning the dragon of chaos: "And there was war in
heaven: Michael and his angels fought against the

dragon; and the dragon fought and his angels. And prevailed not; neither was their place any more in heaven. And the great dragon was cast out, that old serpent, called the Devil, and Satan, which deceiveth the whole world: he was cast out into the earth, and his angels were cast out with him."

Although the idea of Satan as a fallen angel seems not to have been part of the Zoroastrian tradition, it is certainly interesting enough to merit a comment or two. The only clear Old Testament reference to the concept is found in Isaiah 14:12 (incidentally, this is not the same prophet we quoted previously; we are dealing with the first Isaiah this time): "How thou art fallen from heaven, O Lucifer, son of morning!" *Lucifer* means "light-bearer," and it was a name for the morning star, the harbinger of dawn. The Prince of Darkness, in at least one tradition, seems to be a fallen deity of light.

In the first few centuries A.D. Zoroastrian ideas emerged in many different forms. We find them in Mithraism, a religion that flourished especially in the Roman army and which was for a while Christianity's most formidable competitor. The Zoroastrian dualism also appears in the doctrines of the various Gnostic sects and in the religion of Manichaeism, which for a time claimed St. Augustine as a devotee. Unlike the Jews, the Gnostics and Manichees placed as much emphasis on the conflict between Light and Darkness as they did on the Devil, thereby preserving the ancient dualism in a purer form.

Mithraism died a natural death, Gnosticism and Manichaeism more violent ones. Although the latter were quickly condemned by the Church as heresies, it was not until Pope Innocent III pro-

claimed the Albigensian Crusade against the Manichaean Cathari sect of southern France (A.D. 1208)
that the last of these dualistic faiths began to be
eliminated. Even so, a century was to pass before
the Inquisition was able to wipe out the last traces of
the Manichaean heresy. Although Gnostic ideas
can sometimes be found in the writings of later
ages, for example in the prophetic books of the poet
William Blake, neither Gnosticism nor Manichaeism was ever again to flourish as a religion.
The idea of a continuing cosmic war between Light
and Darkness had become a myth; it was no longer
part of any living faith.

Although Zoroastrian influences can be detected
in Eastern religions—for example in Amida, the
Buddha of Light—the dualistic ideas never really
took hold in the Oriental half of the world. The idea
of a cosmic battle between light and darkness apparently didn't fit into their prevailing religious
traditions. To the Hindus, for example, the entire
world, with all of its apparent good and apparent
evil, was only Maya, illusion. Light and darkness
were not opposing entities, only elements in an il-
lusionary world that could be transcended if one
followed the correct spiritual paths. Similarly, the
Chinese, who thought of light and darkness as
complimentary and interpenetrating forces sym-
bolized by yin and yang, would have found it
difficult to see the two principles as elements in a
cosmic conflict.

It is in the West that Zoroastrian ideas have had
their greatest appeal. Judging by the success of Tolkien's trilogy *Lord of the Rings* and the movie *Star
Wars*, that appeal has not faded. Both of these
twentieth-century fictions contain pronounced

Zoroastrian themes. They concern themselves with cosmic battles against evil. The world is presented in black and white, with no intervening shades of gray.

Frodo in *Lord of the Rings* and Luke Skywalker in *Star Wars* must make the decision to participate in the great struggle against the forces of evil. In *Lord of the Rings*, these are personified by the Dark Lord, ruler of the black land of Mordor. In *Star Wars*, the forces of darkness are again led by a Dark Lord, Darth Vader.

It would be interesting to speculate on the reasons why the myth of the battle between light and darkness should be interpreted for us by films and

A scene from the movie *Star Wars*. Darth Vader and Ben Kenobi (Alec Guinness) duel with light sabers. Like the ancient Zoroastrian devil Angra Mainyu, Darth Vader is the personification of evil and the forces of darkness. In *Star Wars* we can see the ancient myth of a cosmic battle between light and darkness. *Courtesy Twentieth Century-Fox. Copyright © 1977 Star Wars Twentieth Century-Fox Film Corp. All rights reserved.*

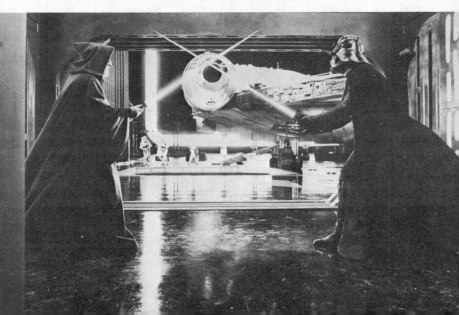

by popular literature when in previous times it was taken up by poets of the stature of Dante and Milton. To delve into that subject, however, would cause us to stray too far from our topic. It is worth noting, however, that whether the story is being told by Milton or Tolkien or movie director George Lucas, we never have any doubt as to the outcome of the battle. Just as Zarathustra was able to foresee the ultimate victory of Ahura Mazdah over Angra Mainyu, we know that Light must win. Whatever vicissitudes Frodo and Luke Skywalker and their companions endure, we know they will prevail in the end. In every version of the Zoroastrian myth, the followers of the Lord of Light must defeat those of the Spirit of Darkness.

Three:
Theories of Light

It would be hard to imagine anything more common, more universal, than light. One would expect it to be relatively easy for science to explain— certainly easier than such forces as magnetism or gravitation. And yet it was not until the twentieth century that scientists were finally able to understand its nature.

Theories about light have always abounded; they go back to the ancient Greeks. Through the nineteenth century, however, all of the theories proposed eventually ran into difficulties. A theory would be advanced saying that light consisted of waves. Then, because that didn't seem to work, it would be replaced by one claiming that light was made up of streams of particles. Something wrong would be discovered with that, also, and a wave theory would again dominate scientific thinking. It wasn't until the advent of the quantum theory in the twentieth century that the problem was finally solved. But in finding the solution to the riddle, the quantum theory introduced new paradoxes of its own.

The road that led to an understanding of light was a long and tortuous one. But along the way so many insights were gained into such things as the colors of the spectrum, the formation of rainbows, the velocity of light, and the nature of infrared and ultraviolet radiation that it is worthwhile to trace that path in some detail.

As far as we know, the first person to propound a theory of light was the Greek philosopher and mystic Pythagoras, who around 540 B.C. said that objects gave off small particles which in some way resembled their source. When these particles struck the eye, they produced vision. A hundred years later Empedocles suggested that the reverse took place, that sight was the result of something which emanated from the eye, striking objects in the field of vision.

Both theories seem to have the same defect: they attempt to explain vision without saying anything about light. They do not tell us why we do not see as well on a moonless night as we do in bright daylight. One can interpret Pythagoras' and Empedocles' theories as implying that light is the stream of particles which emanates from an object or from the eye; but if light is defined that way, it seems to have nothing to do with the illumination that comes from the sun.

Around 430 B.C. Plato, in his dialogue *Timaeus*, attempted to remedy these defects by proposing that vision resulted from the interaction of streams of particles coming from the eye and from daylight. The two coalesced to produce the sensation we call sight. At night the visual stream from the eye is quenched, and we see little or nothing. Plato, then, spoke of two kinds of light which must come together before we can see.

This was a little too complicated for Aristotle's taste, so he attempted to simplify the theory by rejecting the idea that anything emanated from the eye. He said light was caused by the presence of a firelike substance in the air which transmitted some kind of activity from the object. In other words, light was not something that passed through the air, but a substance which permeated the atmosphere during the day and seeped away at night.

Like many of the scientific ideas of the ancient Greeks, these concepts seem strange to us today. If the Greeks had only been willing to perform experiments, they might have come up with something less odd. Unfortunately, they considered manual work degrading, something to be done by slaves, and attempted to understand the universe by means of thought alone. This sometimes had ludicrous results.

In some cases the Greeks arrived at conclusions that were close to being correct. Aristotle, for example, asserted that rainbows were caused by the reflection of light from raindrops. This is partially true. Reflection is involved. The colors of the rainbow, however, are caused by a second process called *refraction*, which was not to be understood until Isaac Newton performed his famous experiments with prisms.

According to a traditional story, in 212 B.C. Archimedes used light to help defend Syracuse against an invading Roman fleet. Archimedes, so the story goes, positioned large concave mirrors to concentrate the sun's rays and set fire to the Roman ships.

If the tale is true, then Archimedes made use of light in a very spectacular way. In 212 B.C. such a device would have seemed even more revolutionary than laser "death rays" do to us today. But is the story true—or is it apocryphal?

Let us look at the contemporary evidence first. The historian Polybius, who wrote at a time when

Archimedes sets the Roman fleet afire with mirrors. Engraving. *Courtesy Bettmann Archive, Inc.*

some of the men who had fought at Syracuse were still alive, makes no mention of burning with mirrors; nor do Livy and Plutarch, who wrote in great detail about Archimedes' role in the defense of

Syracuse. We don't know when the story first came about, but, according to D. L. Simms, a scholar who has studied the available evidence, the earliest known unequivocal statement that Archimedes set fire to the Roman ships is a single sentence written by Anthemius of Tralles 700 years after this feat supposedly took place.

There are other reasons for doubting the story. It is highly improbable that any method of casting or polishing large mirrors existed at the time. Such an instrument could only have been constructed from an assemblage of small flat mirrors. Even if Archimedes had been able to make such a device, no method for calculating the necessary curvature was then known. Finally, it seems unlikely that light from a mirror could have been focused on the damp wood of a ship for a long enough time to set the ship afire, and it could not have been used on the sails, as they would have been furled for battle.

One must conclude, therefore, that the ancients were no more successful in creating technological devices using light than they were in formulating theories. Even though such authors as Euclid, Egyptian astronomer Ptolemy, and others studied and wrote about light, no significant scientific or technological advances were made until the seventeenth century. It is true that lenses were known in classical times, but they seem to have been little more than curiosities. If eighteen hundred years following the death of Archimedes are passed over in silence, little or nothing will be omitted.

The first really significant technical achievement in the field of optics was the invention of the telescope, which is usually attributed to the Dutch optician Hans Lippershey. Around 1608 Lippershey discovered a way to put lenses together in order to magnify distant objects, and thus he constructed the first telescope.

It was the Italian scientist Galileo Galilei, however, who changed the telescope from a mere curiosity into an astronomical instrument. After hearing reports of a magnifying device that had been constructed in Holland, Galileo was able to design a telescope of his own. Using it to explore the night sky, he noted that the moon had a mountainous surface and that the Milky Way was made up of numerous individual stars. Galileo later discovered spots on the sun, and three of Jupiter's moons. So dramatic were his findings that some of his contemporaries refused to look through his telescope, fearing that it was a work of the Devil. When Galileo's astronomical research led to an increasing acceptance of the Copernican system and ultimately to his persecution by the Church, they undoubtedly felt that their suspicions had been confirmed.

It was also Galileo who first attempted to measure the speed of light. The velocity of sound had already been measured and found to be about 1,100 feet per second. Light obviously traveled faster than sound, Galileo reasoned: when a distant piece of artillery was fired, the sound reached the ears a few seconds after the flash was perceived. But it didn't necessarily follow that the transmission of light had to be instantaneous. Galileo decided to conduct an experiment to find out.

He stationed himself and an assistant about a mile apart at night, each with a lantern that could be covered or uncovered at will. Galileo opened his lantern, and the light traveled toward his assistant. When the latter saw the light, he uncovered his

Galileo made one of the earliest attempts to measure the velocity of light. He also constructed one of the first telescopes and used it for astronomical observations. *Courtesy Niels Bohr Library, American Institute of Physics.*

lantern. Galileo was able to measure the time between the opening of his own lantern and the instant at which he could see the light from his assistant's. However, the time span was so minute that the delay could be entirely ascribed to human reaction. Galileo was only able to conclude that light propagated at a velocity that was either "instantaneous or extremely rapid." Today we know that the experiment was doomed to failure from the very first, because light travels so fast that, assuming a one-mile separation, it would make the round trip in about 0.00001 second.

The first experimental determination of the speed of light was made by the Danish astronomer Ole Roemer in 1675. Roemer avoided the difficulty that Galileo had encountered by measuring the time it took light to travel across astronomical distances. Whereas Galileo had tried to time light as it sped across a short distance on earth, Roemer made observations of the moons of Jupiter, which are millions of miles away. The result he obtained—a velocity of about 125,000 miles per second—was later proved too small by a factor of more than 30 percent. He had, however, succeeded in showing that the speed of light, although it was very great, was not infinite.

About fifty years later James Bradley, an English astronomer, made a more accurate determination. Using astronomical observations of a different kind, he concluded that light traveled at a velocity of 188,000 miles per second. The correct figure is 186,000, but since the error was only a little more than 1 percent, this figure represented a significant advance in accuracy.

Despite Galileo's false start, determining the speed of light had turned out to be a relatively simple matter. It was becoming apparent, however, that finding out what light *was* would be much more difficult. Some scientists believed it to be made up of waves, like sound; others that it was composed of particles. These various theories were rarely worked out in any detail; no one even knew what experiments to perform to find the answers that were needed.

The French mathematician and philosopher René Descartes was one of the few to develop a detailed theory of light. According to Descartes light was similar to pressure in a fluid, while colors were analogous to musical tones. None of this seemed unreasonable at the time. However, one of Descartes's hypotheses was that the velocity of light was infinite. This was of course ruled out by the experimental evidence, so the theory eventually had to be abandoned.

In 1678 the Dutch scientist Christian Huygens presented his theory of light to the French academy. Huygens described the manner in which light could be propagated as waves. He explained how light was reflected, and why light rays were refracted, or bent, when they entered transparent substances such as glass or water. Huygens also presented arguments purporting to show that light could not be made up of particles. Only waves cross at right angles without interference; particles would collide with one another. Furthermore, Huygens claimed, only a wave theory could account for the fact that light travels at such great speed.

Although Huygens's concept of light as waves has turned out to be so useful that it is still presented in physics textbooks today, it fared less well among Huygens's contemporaries. As yet, there was no direct evidence of the wave nature of light, and the competing particle theory of Newton seemed to explain the facts better.

There were additional reasons why most scientists preferred the theory advanced by Isaac Newton to that of Huygens. By the time Newton published his book *Opticks* in 1704, he was the most famous scientist in the world. Not only had he invented calculus and discovered the law of gravitation, he had

Like many great scientists, Sir Isaac Newton did his most important work while he was still a young man. The painting below shows Newton as he might have looked when he did his experiments with prisms. *Courtesy Bausch & Lomb Optical Company and Niels Bohr Library, American Institute of Physics.*

also explained such phenomena as tides and comets. In the eyes of many it was impossible for Newton to be wrong.

At the time, Newton's theory did seem to explain reflection and refraction as well as that of Huygens. Huygens had given reasons why he thought a particle theory could not be true. Newton was able to counter these with arguments just as convincing. Today we know that Newton's objections to the wave theory were incorrect. However, at the beginning of the eighteenth century this was anything but obvious.

If light were made up of waves, Newton said, there would be no such things as sharp shadows; light would bend around solid objects. Water waves are observed to do this, he pointed out; and sound bends to so great an extent that it can be heard around corners.

What Newton didn't realize was that light has such a small wavelength that such effects are difficult to observe. For example, the wavelength of green light, which is near the middle of the visible spectrum, is only about one fifty-thousandth of an inch. Contrary to what Newton expected, light does bend around corners. Paradoxically, Newton, who was a careful experimenter, had observed this very effect. But the bending was so slight that it never occurred to him that it could have anything to do with Huygens's waves; he attributed the effect to other causes.

If the scientists of Newton's day had examined his theory in a critical manner, they would have been forced to conclude that there was something very odd about it. Newton's theory said that light had to travel faster in such substances as glass or water than it did in air, in spite of the fact that a solid or a liquid would presumably offer more resistance to the light particles. Furthermore, in order to explain the small fringes that are often seen at the edge of a shadow, Newton had to conclude that matter alternately attracted and repelled light particles, so that rays of light, when they passed the edge of an object, were "bent several times backwards and forwards, with a motion like that of an eel."

Although the particle, or corpuscular, theory was eventually found to be inadequate, Newton did make an important contribution to the understanding of the nature of light. It was he who first correctly accounted for the colors of the spectrum.

In Newton's day it had been known for some time that a prism could split a ray of white light into bands of different colors which always appeared in the same order: red, orange, yellow, green, blue, indigo, violet (admittedly, few of us could distinguish between indigo and violet; however, since the colors of the spectrum were named by Newton, who liked the idea of seven colors, indigo remains). It had always been assumed that the colors were formed by the prism itself, that the glass of the prism somehow acted on white light to give it color.

In order to test this hypothesis, Newton carefully performed some experiments. In one he caused light to pass through two prisms in succession and observed that the second prism caused no further alteration in color. When red light from the first prism passed through the second, the light remained red. Newton found next that the seven colors could be recombined to form the original white light.

Newton realized that this could mean only one

thing: prisms didn't impart color to light at all; they only split it into its different components. Thus far he was correct. When he tried to correlate his results with the particle theory, however, he fell into error.

Newton thought that the various colors were associated with particles of different sizes. The particles of red light were the largest, he said, and those of violet the smallest. Today we know that the various colors are associated with light of different wavelengths. Red light has a wavelength of around 0.00007 centimeter, while that of violet light is about 0.00004. Newton would have been correct if he had only realized that red light consisted of the longest wavelengths rather than the largest particles.

In 1801 the young English physician Dr. Thomas Young gave a lecture before the Royal Society in which he described an experiment which seemed to show that light was made up of waves after all. Under certain conditions, Young said, light can actually cancel itself out. Young called this the *Principle of Interference*.

The phenomenon of interference isn't difficult to understand. Suppose two light waves of the same wavelength are brought together so that the crests of one coincide with the crests of the other. The waves will combine, giving a single wave with crests that are twice as high and troughs that are twice as low. This will result in light that is brighter than that of either of the component waves. This is called *constructive interference*.

But what if the two light waves are brought together so that the crests of one coincide with the troughs of the other? In this case the waves will

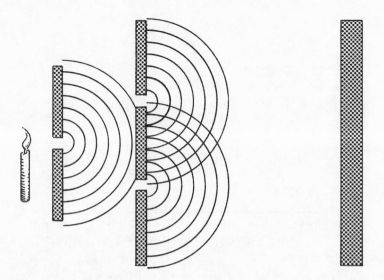

cancel each other out and produce darkness. There is no way that a particle theory can explain this phenomenon of *destructive interference*.

The arrangement of Young's experiment is shown in the accompanying diagram. A light is placed behind a pinhole. The light which passes through it illuminates two other pinholes and then falls on a screen. The light that passes through the second two pinholes adds or cancels, depending on where it comes together on the screen. If the two rays of light are *in phase* (i.e., if crests coincide with crests), a bright spot will be produced. But if they are *out of phase* (if crests align with troughs), the result will be darkness. This setup then should produce a series of light and dark fringes. As Young discovered, this is exactly what happens.

Similar fringes can actually be produced by a single pinhole or slit. If you want to see them, there is a simple experiment you can perform with no equipment at all. Just hold one hand a few inches in front of your eye and look at a light through the space between two fingers pressed tightly together. You will see a number of light and dark fringes. This isn't quite the setup that Young used in his experiment, but the principle is the same.

One would think that Young's experiment would have been thought so conclusive as to establish the validity of the wave theory once and for all. Nothing of the sort happened. By suggesting that light was really made up of waves, Young had cast aspersions on the great name of Newton. The new theory was attacked mercilessly, and Young was ridiculed. The fact that Young was a medical man with little mathematical training certainly didn't help. The scientists of the Royal Society reacted in the way that "experts" always do when their field is invaded by an "amateur": they refused to discuss Young's findings at all, and effectively blocked at that point any further development of the wave theory of light.

Young's theory would most likely have been forgotten if it had not been for work done in France. Even though Newton's theory was generally accepted on the Continent too, the atmosphere there was a little freer. In 1819 the French Academy of Science not only agreed to publish a paper written by Augustin Jean Fresnel on the wave theory of light, it also awarded the paper first prize in its competition for that year.

Unlike Young, Fresnel had a solid mathematical background; hence he was able to work out the wave theory in more detail than Young had and also to explain such phenomena as polarization. Light was made up of transverse waves, Fresnel said. Unlike those of sound, light's vibrations were in a direction perpendicular to the direction of travel. Thus, a ray of light traveling downward from an overhead sun could vibrate in a north-south or east-west direction, or any direction in between. But it could not vibrate up and down. Among other things, this explained why light rays polarized in the same direction could interfere with one another, while rays polarized at right angles to one another could not.

At first none of this made much of an impression in England. In 1834 the British Association's report on the "Progress and the Present State of Physical Optics" commented of the wave theory that "It may be confidently said that it possesses characters which no *false* [their italics] theory ever possessed before." Even though the Young-Fresnel theory successfully explained a vast body of phenomena, the British still clung to the ideas of Newton.

But as the experimental evidence favoring the wave theory increased, even the British were forced to accept it. There is an interesting little story associated with this. British scientists had made what they thought was a convincing objection to the Young-Fresnel theory. Mathematical calculations showed that there should be a bright spot in the center of a shadow cast by a round object, and this was obviously ridiculous. Shortly afterwards it was shown that such a bright spot did exist. Under the proper conditions (e.g., with a small light source shining through a pinhole) it could easily be seen.

Although it had now been established that light consisted of waves, or vibrations, no one could say just what it was that was vibrating. In the case of sound the answer was simple enough: sound waves were caused by vibrations of the air. But what mysterious entity caused the oscillations of light? An understanding of the nature of light seemed as far away as ever.

Only a scientist of the stature of a Newton or an Einstein could have answered that question. Fortunately, the nineteenth century had such a man in James Clerk Maxwell, the great Scottish theoretical physicist. Maxwell is known for his investigations of heat and for his work on the kinetic theory of gases, color and color blindness. His greatest contributions, however, were those contained in his *Treatise on Electricity and Magnetism*, published in 1873. In this book Maxwell not only explained all known electrical and magnetic phenomena; he also proposed that light was electromagnetic radiation, that it consisted of oscillating electric and magnetic fields.

According to Maxwell's theory, if light consisted of electromagnetic radiation, there should be other kinds of radiation with wavelengths shorter and longer than those of light. Maxwell showed that if such waves did exist, their velocity would be the same as that of light; the speed of travel through empty space should not change with wavelength. This was a bold suggestion, for no such waves had ever been detected.

At first Maxwell's ideas encountered resistance from many scientists. But in 1887 Heinrich Hertz demonstrated that electromagnetic waves having all the properties predicted by Maxwell could be produced.

Hertz's waves, which were made with a spark gap and an electrical generator, were in some ways very similar to those of light. They could be refracted through prisms made of pitch; they were reflected from flat surfaces; and like light they were polarized.

James Clerk Maxwell, the great Scottish theoretical physicist. It was Maxwell who showed that light was electromagnetic radiation. *Courtesy Niels Bohr Library, American Institute of Physics.*

In 1895 Marchese Guglielmo Marconi used Hertz's waves to transmit messages over a distance of a mile. Marconi went on to develop "wireless telegraphy." The radiation that Hertz had discovered was what we know today as radio waves.

We know now that there is an entire electromagnetic spectrum, ranging from radio waves to gamma rays. Radio waves have wavelengths that are measured in feet or miles, while the wavelengths of gamma rays are of the order of one hundred-trillionth of an inch. The electromagnetic spectrum is given in the table below. Wavelengths are given in meters. (A meter is about thirty-nine inches, or a little more than a yard.) Numbers are expressed in scientific notation: 10^{-3} is 0.001, while 4×10^{-6} is 0.000004.

The Electromagnetic Spectrum

Name	Wavelength in Meters
Gamma rays	10^{-14} to 10^{-11}
X rays	10^{-11} to 10^{-8}
Ultraviolet	10^{-8} to 4×10^{-7}
Visible light	4×10^{-7} to 7×10^{-7}
Infrared	7×10^{-7} to 10^{-3}
Microwaves	10^{-3} to 1
Radio	Greater than 1

Both radio waves and television use waves in the radio range. Those of television are a few meters in length, while those used by radio stations run to several hundred meters. The best-known application of microwaves is radar, which utilizes waves that are a few centimeters long, roughly one inch.

All of the shorter wavelengths—gamma rays, X rays, and ultraviolet—can in sufficient quantity be harmful to life. Ultraviolet rays are the least dangerous of the three, yet the ultraviolet from the sun would be sufficient to kill all life on earth were it not for the fact that most of it is absorbed by the atmosphere. Even so, enough penetrates to the earth's surface to produce sunburn in anyone who exposes himself to it for too long a time. In moderate quantities, of course, ultraviolet rays are beneficial: it is the ultraviolet in sunlight that produces vitamin D in the skin.

Infrared rays are often thought of as heat rays, since infrared, although not visible to the eye, can be felt as heat. It is not, however, the only kind of electromagnetic radiation that can be converted into heat when it is absorbed by matter. If it were, there would be no such things as microwave ovens.

The applications of the various kinds of electromagnetic radiation in science and technology are so numerous that if an attempt were made to mention them all, there would be room for little else in this book. It should be obvious, however, that Maxwell's theory of electromagnetic radiation has affected our lives as much as any achievement in physics. Furthermore, it is justly considered to be one of the crowning intellectual achievements of the nineteenth century, ranking with Darwin's theory of evolution and Pasteur's germ theory of disease.

As it turned out, however, the electromagnetic theory was not the final word on the nature of light. During the early years of the twentieth century, just

when scientists thought they had learned everything there was to know about light, it was discovered that light sometimes behaved as though it
were made up of particles after all. This led to the
quantum theory of light, which will be the subject
of another chapter.

Four:
Light in Painting

Renaissance interest in light did not begin with Galileo. Long before he performed his experiments, painters were studying lighting in order to give more realistic form to their art.

One striking aspect of most pre-Renaissance art is the absence of light. No attempt is made to give substantiality to forms by modeling them with light and shadow. As a result they remain two-dimensional. Human figures especially seem to lose all their roundness and bulk; they are flat forms drawn on flat surfaces. Examples of this abound in Egyptian art, in Greek vase painting, in medieval manuscript illumination. The styles differ, but the effect is the same.

About the beginning of the Italian Renaissance, however, painters began to feel dissatisfied with the insubstantial spirituality that characterized the art of the Middle Ages. They wanted their work to look more natural. In order to add a third dimension to their paintings, they studied the new laws of mathematical perspective and sought ways to reproduce the effects of light and shadow.

So much has been written about the Renaissance masters' studies of perspective that one can easily get the impression that this was their only important technical discovery. But of course it wasn't, as

Leonardo da Vinci points out in a passage from his notebooks:

> The first marvel of painting is that it appears detached from the wall or other flat surface, deceiving people of subtle judgment with this object that it is not separated from the wall's surface. . . and this is why the painter must make it his concern to study shadows which are the companions of light. . . . . The second thing is that the painter must, through serious reasoning and subtle investigation, fix the true quality of light and shade. . . . The third thing is perspective, which is a most subtle discovery in mathematical studies, for by means of lines it causes to appear distant that which is near, and large that which is small.

In other words, if an artist is to create the illusion of reality, he must know how to use *chiaroscuro*. Only after he has mastered that should he begin worrying about perspective.

The word *chiaroscuro* comes from the Italian terms *chiaro* ("light" or "bright") and *oscuro* ("dark") and denotes the modeling of forms by the use of light and shadow. Chiaroscuro had been discovered by Greek painters in Hellenistic times and had also been used by Roman artists, but it had been forgotten during the Middle Ages. Hence, the Renaissance masters had to rediscover it before they could attain the realism they desired.

Artists were not, however, to become preoccupied with light as an independent pictorial entity until much later. Although some artists continued to place great emphasis on the use of light to create atmospheric effects, it wasn't until the advent of

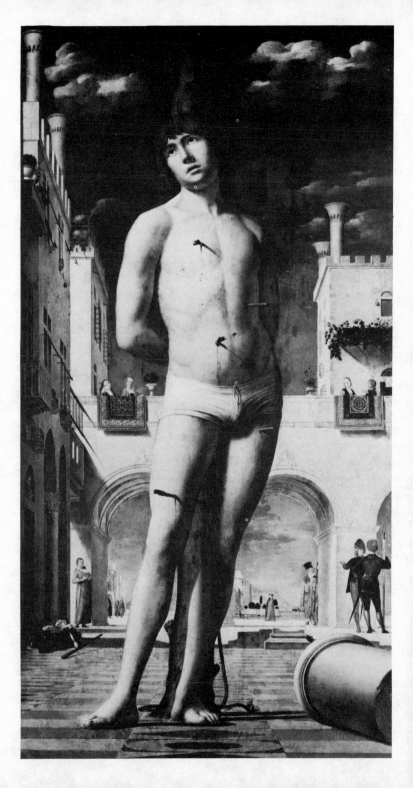

impressionism in the nineteenth century that artists began to attempt to paint light itself.

Anyone who looks at much Renaissance art soon becomes aware of a certain sameness about the lighting. Almost without exception, it is the diffuse light that might enter through the window of a studio or of a church. In those days painting was an indoor activity. The grinding and preparation of pigments was a tedious business, and in many cases paintings were executed in fresco directly on walls. The collapsible paint tube was an invention that lay hundreds of years in the future, and canvases were often so large that it would have been impossible to carry them around outdoors. The light that artists painted was the light they saw while they were working, and this was never direct sunlight.

Around the end of the sixteenth century an Italian artist named Michelangelo Merisi da Caravaggio discovered that lighting could be used to create a great sense of drama. Strongly directed artificial light or a shaft of sunlight coming through a window could be used to highlight figures and give emphasis to the primary subject. If at the same time other parts of the picture were in deep shadow, compelling effects could be produced.

Caravaggio's contemporaries were stunned by his work. Before long, Caravaggio had numerous followers, called Caravaggisti, and his influence spread all over Europe. Even sculptors and architects attempted to find ways to produce effects of light and

The modeling of forms with light and shadow is called *chiaroscuro*. The painting reproduced opposite is *St. Sebastian* by Antonello da Messina. *Courtesy Staatliche Kunstsammlungen Dresden.*

dark that were Caravaggesque. Caravaggio has to be considered one of the most influential artists in the history of painting.

Today it is Rembrandt who is perhaps best known for his use of dramatic chiaroscuro effects; yet without the example of Caravaggio, Rembrandt's art could have been very different. Rembrandt would certainly have still become a great artist, but most likely he would not have developed the style we associate with him.

Rembrandt van Rijn became interested in strong chiaroscuro effects while he was still a young man. He used these throughout his life, making chiaroscuro even more expressive than Caravaggio had ever thought possible. Rembrandt didn't allow his preoccupation with light and dark accents to become limiting, however. His paintings, especially the later ones, have a richness of color and a plasticity of form that create effects more profound than any that could be attained by the use of light alone.

One of Rembrandt's most famous paintings, the so-called *Night Watch*, is notable for its bold use of chiaroscuro effects. However, *The Night Watch* was not its original title, and Rembrandt never intended it to be a night scene. When the painting was cleaned in 1946–47, it became apparent that the nocturnal appearance had been caused by the darkening of the layers of varnish that had been applied over the paint. Rembrandt had actually intended to represent a scene in the late afternoon. Perhaps it should be considered a tribute to Rembrandt's great skill that this painting should have made a deep impression on so many people, even though it had darkened to such an extent that they could no

Rembrandt's *Night Watch*. Rembrandt was noted for his dramatic chiaroscuro effects. *Courtesy Rijksmuseum, Amsterdam.*

longer tell whether the event it depicted had taken place at night or during the day.

After Rembrandt there are no important artists who are remembered for their use of chiaroscuro. Rembrandt had used the effect so successfully that succeeding painters felt they must travel different roads if they were to be more than imitators. Eventually some of them turned from the expressive use of light in painting to attempts to study the nature of light itself.

Of all the pre-impressionist painters, the one who came nearest to achieving this was Turner. However, before we discuss Turner's work, it might be well to say a few words about William Blake.

Art critics are never able to fit Blake smoothly into their discussions of the development of artistic trends and traditions. Blake was so original and so

William Blake, *Adam and Eve Sleeping*. This watercolor by Blake is infused with a light that has no obvious source and thus gives a visionary quality to the painting. *Courtesy Museum of Fine Arts, Boston.*

eccentric that he seems to stand apart from any artistic tradition.

Some of Blake's contemporaries thought him mad. Although today we like to think that the ability to create great art is in itself a proof of sanity, we have to admit that Blake was somewhat peculiar, at the very least. While he was still quite young, he began to show a prophetic turn of mind and to see visions. When he painted, Blake insisted on creating from his own imagination; he refused to paint from life.

Blake had no desire to make use of chiaroscuro, which he called that "infernal machine." Yet in his own way he infused a sense of light into his work. The only way to describe this effect is to call it a "visionary light." This light, which permeates all of Blake's paintings, seems to have no source, it does not fall from one direction or another, but seems present everywhere. Blake often combined his art with his poetry, and the illustrations in his prophetic books are in some ways reminiscent of medieval manuscript illumination.

Joseph Mallord William Turner is a little easier to discuss, even though his work was as nearly unique as that of Blake. Turner, who had an early success as a landscape painter, became more and more interested in finding ways to depict light itself. As he matured, his style became increasingly free and abstract. In spite of violent criticism, Turner persisted in his quest. In his later work, recognizable forms disappear almost entirely; there is nothing left but light and space and the natural elements.

Unlike the Renaissance artists and Caravaggio and Rembrandt, Turner was not interested in the effects of light and shadow. And unlike the im-

J. M. W. Turner, *The Slave Ship*. In Turner's later paintings, forms become insubstantial, so that there is little left but the interplay of light and space with the natural elements. *Courtesy Museum of Fine Arts, Boston.*

pressionists, although he is sometimes considered their forerunner, he was not interested in analyzing the appearance of light in nature. In the paintings of Turner, light becomes a romantic symbol. Perhaps one can best describe this light by saying that it appears to represent some "cosmic force." In Turner, light is always an effect, and his work can be very impressive. However, at some point his "light" ceases to be light at all. Instead, it is a symbol for the often slumbering, sometimes vengeful forces of the universe.

It was Turner's romanticism which prevented him from penetrating the secrets of light the way the impressionists were to do later. The impressionists, who were interested not in the dramatic but in the commonplace, who painted "boring" landscape after boring landscape, would find a way to depict light via the medium of canvas and oil.

In the spring of 1874 a group of young avant-garde painters—among them Claude Monet, Pierre-Auguste Renoir, Camille Pissarro, Alfred Sisley, Edgar Degas, Paul Cézanne and Berthe Morisot—organized an exhibition of their work in Paris. Their independence and the revolutionary nature of their art had led to repeated conflicts with the official Salon. They felt, therefore, that it was necessary to hold an independent exhibition in order to bring their work before the public.

In order to understand their motives, it is helpful to know something about the state of French art in the nineteenth century. For decades French art had been dominated by the conservative Academy of Fine Arts which had replaced the old Royal Academy of Painting and Sculpture in 1796. Through its Salon the Academy patronized artists who accepted conventional academic standards. So great was the Academy's power that it was able to make art of which it disapproved virtually unsalable. This power was largely the result of the nature of the buying public. After the French Revolution of 1789 the market for paintings had shifted from a small group of aristocratic patrons to the less knowledgeable bourgeois public. Since the bourgeoisie were not so confident of their artistic tastes, they eagerly accepted any art which bore the academic stamp of approval.

The Academy's annual Salon was actually a huge marketplace in which up to 5,000 paintings were viewed and purchased. Crowded together as closely as their frames would permit, they filled the walls from eye level to ceiling. Every year the Salon jury rejected a number of works, sometimes as many as 4,000. These were often stamped on the back with an R for *refusé*, and since few bourgeois patrons would purchase a work that had been officially declared to be "defective" by the government, a painter who had been repeatedly rejected might find that he had to choose between giving up art and starving.

Since the academicians were so concerned with "upholding standards," paintings which showed great originality were very rarely accepted. The Salon juries much preferred work that showed technical finish but did nothing to upset the accepted classical traditions. When a daringly original artist such as Édouard Manet did somehow manage to have his work exhibited in the Salon, the academicians were likely to compensate for the "mistake" by becoming more restrictive the following year.

The artists who organized the first impressionist exhibition in 1874 were not entirely unknown to the Salon public. A number of them—including Monet, Degas, Renoir and Pissarro—had exhibited there several times. However, much of their best work had been rejected. Since a number of private collectors were beginning to buy their work, these artists felt that an exhibition would not only increase their sales but would also give the public an opportunity to see those of their paintings that were unacceptable to the Academy.

They did not yet refer to themselves as impressionists, but simply as the *Société anonyme des artistes peintres, sculpteurs, graveurs, etc.* The exhibitors were a mixed group. Whereas some of them wanted to declare aesthetic war, others were still convinced that exhibiting in the Salon was the only real road to success. Indeed, Manet, who had by this time become something of a symbol of the revolutionary in art, was so convinced of this that he refused to have anything to do with the project. The exhibitors themselves weren't even agreed that the exhibition was necessary. Degas, for example, felt that his career might be harmed by association with the group. Nevertheless, once the showing had been planned, he joined in and did everything he could to ensure its success.

Once a site had been found and the list of contributors agreed on, the committee in charge somewhat lazily left Renoir to make most of the final arrangements. Renoir's brother Edmond was given the frustrating task of editing the catalog. Among other annoyances, Edmond had difficulties with Degas because his paintings were ready only at the last minute. And although he delivered his canvases in plenty of time, Monet was no less irritating in the monotony of his titles: *Entrance of a Village, Leaving the Village, Morning in a Village,* and so on. When the exasperated Edmond objected to the title of a view of Le Havre, Monet told him, "Why don't you just put *Impression!*" The painting was duly cataloged as *Impression, Sunrise.*

When the exhibition opened, those critics who did not ignore it attacked it violently, and the public came only to laugh. One critic, Louis Leroy, a writer for the satirical paper *Charivari,* happened to notice Monet's painting of Le Havre. Taking his cue from the title, he sarcastically played on different

Monet, *Impression, Sunrise*. This painting, one of those attacked by the critics who attended the first impressionist exhibition in 1874, gave the movement its name. *Courtesy Musée Marmottan, Paris.*

variations of the word *impression* as he wrote his review, which he called "Exhibition of the Impressionists."

Leroy was not the only one to laugh. Someone invented a joke saying that the method of these painters consisted of loading a pistol with paint and firing it at a canvas. Another visitor called the exhibition a "laughable collection of absurdities." Yet another remarked that Cézanne must have worked while in the midst of an attack of delirium tremens. One of Pissarro's paintings was said to be composed of "palette-scrapings placed uniformly on a dirty canvas," while a work of Monet's was compared to "wallpaper in its embryonic state." Not all of the critics were so imaginative; some could do no better than to refer to the impressionists as "madmen."

Between 1876 and 1886 seven more impressionist exhibitions were held. The criticism continued; the impressionists were even attacked as "communists," although none of them had had anything to do with the Paris Commune of 1871. In all fairness, though, most of the critics focused on the artistic qualities of the work, rather than on unfounded political charges. "Try to explain to M. Renoir," said the critic Albert Wolff in *Le Figaro*, "that a woman's torso is not a mass of flesh in the process of decomposition with green and violet spots which denote the state of complete putrefaction of a corpse!" The impressionists were characterized as "unfortunate creatures stricken with the mania of ambition," as demented as the inmates of Bedlam.

Even by 1893, nearly twenty years after the first exhibition had taken place, the cries of horror had not completely died down. When a prominent collector died, leaving sixty-five impressionist paintings to the French government, the objection was raised that "For the government to accept such filth, there would have to be great moral slackening"; and several academicians threatened to resign from the *Ecole des Beaux-Arts* in protest. The state finally did accept the bequest, but only after refusing eight paintings by Monet, eleven by Pissarro, two by Renoir, three by Sisley, two by Manet, and three by Cézanne. Only Degas saw all of his works accepted.

Today when we look at a landscape by Monet or

at Renoir's nudes or his attractive young women, we wonder why impressionist painting provoked such violent reactions. After all, the impressionists were doing nothing more than trying to recreate the effects of light in the open air; they were only attempting to paint what they saw.

In order to understand why the public and the critics reacted as they did, it is necessary to know something about the accepted academic technique. Although impressionist work doesn't look startling to us, to its contemporaries it hardly seemed to be art, because it looked nothing like the work that was officially promoted.

The typical academic painting of the era was notable for its "finish." The painter normally did not allow his brushstrokes to be visible in the completed work, and he meticulously smoothed out the boundaries between different colors so that there would be no sharp juxtapositions. Academic painters were masters of chiaroscuro. They posed their models in the studio, where light came in through a window and created a gradual transition from light to shade. They modeled their forms with even shadows of gray and black in order to give the impression of roundness and solidity.

Even when doing landscapes, they painted in the studio, after sketching their subjects in the open. They gave each object its own even color. A leaf was green, a face flesh-colored; drapery was blue or yellow or pink. They had forgotten the observation of Leonardo that "Green and blue are invariably accentuated in the half shadows, yellow and red and white in the highlights." They knew nothing of Leonardo's advice to painters about depicting a scene in the outdoors:

Therefore, painter, show in your portraits how the reflection of the garments' colors tints the neighboring flesh. You desire to depict a white body, surrounded by air alone. White is no color of itself; it changes and adopts part of the colors around it. If you see a woman clad in white, in an open landscape, she will be of such brilliance on that side towards the sun as to dazzle the eyes almost as much as the sun itself. That side of her, however, which is illuminated by the light from the sky will have a bluish hue. Should she stand near a meadow between the sunlit grass and the sun itself, the folds of her gown on which the light of the meadow falls will show the reflected light of the green meadow.

The academic painters ignored such admonitions. They painted not what they saw, but what they had been taught to see. They never noticed that the leaf of a tree looks bluish when illuminated from the front and yellowish when the sun shines through it. They had no idea that a meadow can be one color when one views it with the sun at one's back, and a slightly different tint when seen from the opposite direction. They didn't know that colors often mix in nature so that, for example, the blossoms of apple and pear trees normally appear dirty white—an effect arising from the combination of pink and white petals, green leaves, and red or yellow stamens. They didn't realize that shadows on the snow are not gray.

The impressionists, on the other hand, presented light and color in an entirely different way. Realizing that shadows are not brown or black but are colored by their surroundings, they painted snow with blue shadows and put violet tints in skin and clothing. They ignored the so-called local colors of objects and painted them in hues that showed the influence of atmospheric conditions. In order to better represent the scintillation of light, they painted in bright colors which were to be merged in the eye when the painting was viewed from a dis-

Renoir, *The Luncheon of the Boating Party*. Although it was Monet who explored the possibilities of impressionism to the fullest, it is perhaps Renoir's work that we enjoy the most. *Luncheon of the Boating Party,* with its bright colors, its light and warmth, its airy mood, is quintessential Renoir. *Courtesy The Phillips Collection, Washington, D.C.*

tance. They applied paint in perceptible brushstrokes, blurring the outlines of objects. Working rapidly in order to capture the immediacy of a visual sensation on canvas, they created forms which seemed to lose their solidity. It was as though they were trying to paint not objects, but the light in which the objects were seen.

To the academicians and to the public and the critics of the day, it seemed as though the impressionists were intent on mocking everything they had come to know as art. Consequently, the impressionists were greeted with outrage and derision. Although the public couldn't shake the convictions of the impressionists, it could make life difficult for them simply by refusing to purchase their paintings. As a result the impressionists who were not, like Degas, independently wealthy had to endure years of poverty while they struggled to gain acceptance. Often they lacked the money even to buy paints. They were hounded by creditors and evicted from their lodgings. Sometimes they quite literally starved; for example, malnutrition seems to have been a contributing factor in the death of Monet's wife.

And in the end they suffered the fate of all successful revolutionaries. They succeeded so completely that today their work no longer seems startling at all. After a century it sometimes looks only "pretty"—almost innocuous.

Nowadays, although it is customary to pay homage to the impressionists and to admire their attempts to recreate the effects of light, we often find ourselves becoming more intrigued by the work of their successors, the so-called post-impressionists such as Van Gogh, Gauguin and Toulouse-Lautrec (all three of whom, incidentally, began as impressionists). However, if it had not been for impressionism and its liberating influence, post-impressionism would not have been possible. There would have been no Gauguin with his areas of vivid color, no Van Gogh with his expressionistic brushstrokes. Lautrec would not have found the unhampered vision he needed to portray his singers, circus performers, actors and prostitutes. It was the impressionist revolution that freed painting from the suffocating domination of the Academy and made possible the trends which finally led to what we know as modern art.

Monet, possibly the greatest impressionist painter of them all and certainly the one who most fully explored the possibilities of the style, did more than pave the way for the *fin de siècle* painters. Monet, who outlived Van Gogh, Gauguin and Lautrec by more than twenty years, eventually found a way to dissolve structure so extensively that he is sometimes hailed as a precursor of abstract expressionism.

Monet was so fascinated by light and color that they became an obsession to him, as he discovered tragically early one morning in 1879 after he had spent the night by the bedside of his ill young wife, Camille. The morning light revealed that she was dead. To his horror Monet found himself carefully examining the changing colors of death in her face. As the light grew stronger and Camille's pallor deepened, he noted the gray, blue and yellow tonalities of death. Realizing that he had become so much a prisoner of his fascination with visual experience that, even in the midst of personal tragedy, he could think of nothing else, he feared that art was costing him his humanity. Describing the experience later in a letter, he compared himself to an animal condemned to turn a millstone.

Nevertheless, he continued experimenting. In 1890 he settled at Giverney near Paris on the river Epte. There he painted from a boat in which he had cut grooves to hold canvases. Monet would work on one painting until he noticed a change in the light. Then he would put that canvas away and take up another, continuing to paint the same subject as it appeared in the changed light. Years later, while speaking of the famous "Poplars" series that he painted from the boat, he commented that the effect shown in one of the paintings had lasted only seven minutes, until the sunlight left a certain leaf. He once said that he wished he had been born blind and then had suddenly gained his sight so that he could paint without being able to identify the objects he saw before him.

In 1890 Monet began work on his most famous series of paintings. He did fifteen canvases representing two haystacks at various hours of the day. He began by planting himself before the haystacks early one morning with a supply of blank canvases placed beside him. As the sun rose, he replaced each of these with a fresh one at hourly intervals. The following day, he placed himself in the same spot and worked on one canvas after another as before. This went on day after day until the paintings were finished.

Next Monet repeated the exercise by painting the façade of the cathedral at Rouen, this time in an even more elaborate way. He painted in different kinds of light and at different seasons of the year. The results were even more successful. In these

Monet, *Rouen Cathedral: Tour d'Albane, Early Morning*. In Monet's later paintings, forms lose all their substantiality; everything is subordinated to the artist's obsession with the interplay of light. *Courtesy Museum of Fine Arts, Boston*.

paintings the cathedral seems to lose all of its sub-stantiality and to dissolve into a complicated inter-play of light.

Monet's last series of paintings, completed late in his life after he had had several operations for cataracts and was nearly blind, express the alchemy of light to the greatest degree. In these paintings of a pool with water lilies, form disappears altogether, so that nothing is left but abstract relationships of light and color. Monet has at last forsaken the world of material objects and is painting pure light. It is in this series of paintings that he is said to have antici-pated the work done by the abstract expressionists many years later.

Monet died in December 1926 at the age of eighty-six. "I have always had a horror of theories," he said in a letter written shortly before his death. "I have only the merit of having painted directly from nature, trying to convey my impressions in the presence of the most fugitive effects." With his "horror of theories" Monet left no writings in which he attempted to explain what he was trying to do. We have only his paintings. Numerous books and treatises have been written about them, but perhaps a single sentence by Cézanne best sums up Monet's art. "Monet is only an eye," Cézanne said, "but what an eye!"

Five:
The Invention of Photography

One day in 1827, after he had completed a lecture, the distinguished French chemist Jean Dumas was approached by a woman who seemed very agitated. Introducing herself as the wife of the painter L. J. M. Daguerre, she told Dumas that her husband seemed obsessed with the idea of producing permanent pictures from the images in a camera obscura. Her husband was so preoccupied with his research that he would lock himself up in his studio for days at a time, not sleeping, and eating without taking any notice of the food that was brought him. Was he mad? Mme Daguerre asked. Or did Dumas think that what her husband was attempting to do was really possible?

As a scientist Dumas was aware that certain chemicals were sensitive to light. Hence he was reluctant to tell Mme Daguerre that her spouse was seeking the impossible. "In the present state of our knowledge," he replied, "it cannot be done; but I cannot say that it will always remain impossible, nor set the man down as mad who seeks to do it."

At least that is how Dumas reported the conversation years later, after both Daguerre and the daguerreotype had become famous. Since he undoubtedly wanted to present himself in the best possible light, Dumas may have altered a word or two, either consciously or unconsciously. If so, the

distortions are probably not serious ones; the conversation as Dumas reported it does have a ring of
truth. Many of Daguerre's contemporaries would

Louis Jacques Mandé Daguerre, inventor of the first
practical photographic process. This 1844 daguerreotype
was made by J. B. Sabatier-Blot. *Courtesy International
Museum of Photography, George Eastman House,
Rochester.*

have thought him mad if they had known what he was attempting. But Dumas would have been very much aware that light-sensitive chemicals existed and that numerous experiments using them had been carried out.

Mad or not, Daguerre succeeded. Twelve years later, in 1839, he presented his discoveries to the French Academy of Sciences; the daguerreotype had become the first practical photographic process. Within a short time it had become a sensation, the most talked-of topic in Europe. Instantly Daguerre became world-famous. In some quarters photography was called the "greatest invention since the printing press," but in others it was criticized as blasphemous. "God created man in His own image," said a writer in the German publication *Leipziger Stadtanzeiger*, "and no man-made machine may fix the image of God. Is it possible that God should have abandoned His eternal principles, and allowed a Frenchman in Paris to give to the world an invention of the Devil?"

Although Louis Jacques Mandé Daguerre developed the first workable photographic process, he was not the sole inventor of photography. Another Frenchman, Joseph Nicéphore Niépce, made the world's first permanent photograph; and William Henry Fox Talbot, an Englishman, developed the negative-positive process that we use today. However, it would not be accurate to call Niépce or Talbot the inventor of photography either. The development of photography spanned more than a century, and it was the work of a number of people, each of whom deserves part of the credit.

When Daguerre gave his process to the French government in 1839, cameras had already been used to make pictures for hundreds of years. They were not permanent pictures, however; the first ones were nothing more than scenes projected on the walls of darkened rooms.

The word *camera* comes from the It.' 'n term *camera obscura*, "darkened room." The prin˪ ˙ ˦f the camera obscura is described in Leonaɪ notebooks. If a small hole is made in the wall ot ˎ very dark room and the rays of light emanating from the hole fall on a piece of white paper, then, says Leonardo, one will see images of objects that are outdoors. The images will be smaller than the objects, and they will be upside down, but they will retain the original shapes and colors.

The invention of the camera obscura has been ascribed to such people as Roger Bacon, the Italian architect Alberti, Leonardo, and the Italian sculptor della Porta. But in fact it was known as early as the tenth century; Alhazen, the Arab scholar and writer on optics, speaks of using such a device to observe the eclipse of the sun.

At first the camera was simply a darkened room big enough for a human being to enter. By the seventeenth century portable cameras had been developed, and by the early nineteenth century they were quite common and were frequently used by artists as an aid in perspective drawing. Some were now equipped with lenses that projected images onto plates of ground glass. Artists could trace drawings of natural scenes by placing sheets of thin paper on the glass. They found the device to be so fascinating that many of them gave no heed when Sir Joshua Reynolds warned them that dependence on a camera could stifle an artist's imagination.

Although some of the cameras were quite sophisticated in design, there was not yet any such thing as photography. To make a photograph, there are three requirements. The first is a camera with which to create an image. The second is a light-sensitive material of some sort on which the image can be recorded. Finally, some way must be found to make this image permanent. It was the latter which proved to be the major stumbling block in the development of photography.

The first person who clearly recognized that the action of light could cause changes in certain substances was Johann Heinrich Schulze, a professor of anatomy at the University of Altdorf. In 1725, while trying to make phosphorus, Schulze filled a flask with a mixture of nitric acid and chalk which happened to contain a little silver. When the mixture was placed in the sunlight, the color of the side facing the sun changed from white to purple. When Schulze exposed a similar mixture to the heat of a fire, there was no change in color; he therefore attributed the color change to the action of light.

Continuing his researches, Schulze added a greater amount of silver to his mixture, and discovered that the changes then took place more rapidly. He substituted other substances for the chalk and thus established that the important ingredient was the silver in the form of silver nitrate, a product of the combination of silver and nitric acid. Experimenting further, he learned that the silver solutions turned black when spread on skin, wood or bone and exposed to light.

Although Schulze believed that his findings were of considerable importance, he made no attempt to develop any practical application for the process. But we know now that it was his discovery that led to the invention of photography. With a few minor exceptions, all photographic processes have been based on the use of silver; all modern photographic films make use of the sensitivity of silver compounds to light. In fact, the black in a black and white negative or print is actually silver, although it looks dark because it is made up of finely divided particles. Color films make use of silver, too; but here the silver is replaced by dyes in the final slide, negative or print.

Schulze wasn't able to explain why light had a darkening effect on silver compounds; this was not to be understood until long after photography had become a workable process. Only after the development of quantum mechanics in the twentieth century was it discovered that the creation of a photographic image was a complicated process involving the absorption of light energy by electrons.

Fortunately, it is not always necessary to understand a process in order to work out practical applications for it. Contrary to popular belief, technology does not consist solely of the application of scientific knowledge. Sometimes the technological process comes first and the scientific explanation afterwards. For example, steel was produced long before science had any adequate understanding of the properties of metallic alloys; and when the Wright brothers built their airplane, most physicists were convinced that heavier-than-air flight was impossible. It should not be surprising, therefore, that photography came into existence approximately a hundred years before scientists were able to explain the formation of the photographic image.

Around the beginning of the nineteenth century, Thomas Wedgwood, the son of Josiah Wedgwood, the famous English pottery manufacturer, attempted fo fix the images produced by a camera by spreading a solution of silver nitrate on paper or white leather. His experiments, however, were unsuccessful; the light passing through the lens was too dim to create a discernible image. So he concentrated instead on making white-against-black silhouettes by placing objects such as leaves or the wings of insects on the sensitized paper and exposing it to the sun. Since sunlight will not pass through an opaque object, this produced a white shadow on a black background.

These sun prints, however, were not permanent. They had to be kept in the dark to prevent the white parts of the pictures from darkening also, which would cause the images to quickly disappear. Wedgwood never found a solution to the problem; he died at the age of thirty-four, only three years after the publication of his experiments. It was not until some years later that an adequate fixing agent was found.

Wedgwood's researches were continued by the great English chemist Sir Humphry Davy. But although Davy did discover that silver chloride was more light-sensitive than silver nitrate, he too failed in his attempts to produce images in a camera. Nor was he able to find a fixing agent. "No attempts that have been made to prevent the uncoloured parts . . . from being acted upon by light, have as yet been successful," he commented in 1802 in the *Journal of the Royal Institution* when reporting on his own and his friend Wedgwood's work.

The problems associated with using silver compounds proved to be so formidable that the first permanent photograph was not made with silver but with a kind of asphalt called bitumen of Judea. This first picture was made by Joseph Nicéphore Niépce (pronounced *nyeps*) in 1826, thirteen years before

Joseph Nicéphore Niépce. Niépce made the world's first photograph, but his process never became feasible. It was Niépce's partner Daguerre who was to introduce a workable photographic system six years after Niépce's death. *Courtesy Kodak Museum, Harrow.*

Daguerre announced the success of his process based on silver.

Niépce, who had been engaged in photographic researches since 1816, abandoned his attempts to produce images with silver because, like Wedgwood and Davy, he had been unable to find a fixing agent. Experimenting with bitumen of Judea, he discovered that when a plate covered with the bitumen was exposed to light in a camera, the exposed parts of the plate hardened, while the rest of the picture remained soft and soluble, so that the bitumen could be washed away.

In order to make his photograph, Niépce dissolved the asphalt in oil of lavender and spread a thin layer of the solution onto a metal plate. After exposing the plate, he washed the plate with more lavender oil. The asphalt that had been exposed to light remained, with the image visible, while the unexposed portion was washed away, leaving the bare metal.

Bitumen of Judea was not any more sensitive to light than the substances Wedgwood and Davy had used. But Niépce simply exposed the plate for a longer period—eight hours. His process thus was hardly practical for portraits, and when used to photograph views from nature, it gave rather odd-looking results. Niépce's first successful photograph, taken from a window of his house, was of a courtyard; the exposure was so lengthy, however, that the sun appeared to be shining on both sides of the picture. Moreover, the image was indistinct. This photograph, now in the Gernsheim collection at the University of Texas, can be clearly made out only if it is held at a particular angle against the

The world's first photograph, made by Joseph Nicéphore Niépce in 1826. Because an eight-hour exposure was necessary, the sun seems to be shining on both sides of the courtyard. The reproduction exaggerates the surface irregularities of the plate; in the original the sky appears as a white blank, and the buildings do not seem so mottled. *Courtesy Humanities Research Center, University of Texas.*

light; the plate is as shiny as a mirror, and the image is very faint.

In January 1826 Niépce received a letter from Daguerre, at that time a stranger to him. Daguerre wrote that he had been given Niépce's address by their mutual lens maker. He had heard of Niépce's experiments, he said, and added that he was working along similar lines.

Niépce had no intention of disclosing any of his secrets, so he sent a vague reply. When the correspondence was resumed more than a year later,

Niépce and Daguerre were equally suspicious of each other. Daguerre sent Niépce something called a "smoke drawing" which had little or nothing to do with his researches. In response Niépce sent, not a photograph, but a lightly etched plate which he told the engraver Lemaître, "could in no way compromise the secret of my discovery." Nevertheless, the two continued the correspondence and about the middle of 1827 Niépce suggested that he and Daguerre become collaborators. Nothing came of it.

The correspondence was reopened in 1829, and Niépce and Daguerre signed partnership papers later that year. The partnership turned out to be less than fruitful. Niépce died four years later, and Daguerre went on to develop his own photographic process; there seemed to be little possibility of making Niépce's discovery into anything practical.

In 1837 Daguerre succeeded in making a photograph of a corner of his studio. After secretly demonstrating his method to François Arago, director of the Paris observatory, he lectured on it to the Academy of Sciences. A few months later a bill was passed by the French Parliament which granted an annuity of 6,000 francs to Daguerre and an additional 4,000 francs to Niépce's son Isidore, who had taken his father's place as Daguerre's partner. In return for the money, Daguerre divulged all the details of his process. Because it was being purchased by the government, theoretically anyone was free to make use of it. Later, however, Daguerre patented it in England.

The success of Daguerre's process, called the daguerreotype, can be attributed to what was probably the most significant discovery in the history of photography: how to develop a latent image.

All previous attempts at making photographs with silver compounds had made use of the fact that these substances gradually darkened when they were exposed to light. However, exposure times, if an image could be obtained at all, were impossibly long. Daguerre discovered that an image too faint to be seen by the eye could be intensified by the process of development.

Like hundreds of other important discoveries, this one was accidental. Daguerre had put an unsuccessfully exposed plate into a cupboard, probably intending to reuse it. When he took the plate out again after several days, he was astonished to discover that it now showed a distinct image. Surmising that the development had been caused by the action of one of the chemicals stored in the cupboard, he set to work to discover by trial and error which one it was. Finally he established that the chemical in question had been mercury from a broken thermometer.

Before long Daguerre was developing his photographs by placing his plates in the vapor of heated mercury. He was thus able to improve considerably on the eight-hour exposures which Niépce's process required. To make a daguerreotype, only twenty to thirty minutes of exposure to light was needed.

Daguerre's photographs were still not permanent. However, that problem too was solved when Daguerre discovered that an image could be fixed by bathing the developed plate in a solution of hot salt water. Later, following a suggestion of the English astronomer Sir John Herschel, he substituted

Sir John Herschel. One of the best-known photographs by the first great portrait photographer, Julia Margaret Cameron. It was Herschel who discovered the process which Daguerre and Talbot used to fix their photographs, a process still used today. *Courtesy International Museum of Photography, George Eastman House, Rochester.*

a substance known as "hyposulfite of soda," which is still the most commonly used fixing agent today. Although the modern name of the compound is sodium thiosulfate, photographers still refer to it as "hypo."

By modern standards, Daguerre's process was a cumbersome one. A plate of silver-plated copper was polished with pumice powder and oil; then it was dipped into a solution of nitric acid, taken out, heated over the flame of a lamp, and rinsed in nitric acid again. The next step had to be performed in the dark. The plate was held in the fumes of heated iodine until a coating of silver iodide had formed. Then it could be put into a camera. After an exposure of five to forty minutes, depending on light conditions, the plate was removed and developed in mercury vapor. If it was now viewed from the proper angle, one could see a positive image; from other angles it had the appearance of a negative.

This method of making photographs was not only exacting and tedious; it could also be dangerous. Two of the substances used, iodine and mercury, are poisonous. The latter is especially dangerous, as it can accumulate in the body, and its effects appear only after a period of time. (The expression "mad as a hatter," incidentally, is a reference to the fact that ingestion of mercury, which hatters used in their craft, can cause insanity.)

Besides the risks involved, there were other disadvantages. Daguerreotypes showed a left-to-right reversal; i.e., they were mirror images. Because the images were easily marred, finished plates had to be kept under glass. Exposures were long, and posing could be torture. Attempted street scenes invariably turned out to be curiously depopulated; pedestrians and vehicles did not stay in one place long enough to register on the plate. Daguerreotypes were uncomfortable to view because of the metallic glare. Finally, since there were no negatives, a picture could not be duplicated; it could only be rephotographed or copied by hand.

Nevertheless, the daguerreotype captured the imagination of the public. By the end of 1839 the mania had seized all of Paris, and it soon spread throughout the world. Within five months after Daguerre had written a seventy-nine-page booklet describing the process, it had appeared in more than thirty editions, translations and summaries.

But daguerreotypes were to be made for only about a decade. In fact, when Daguerre announced his process, it was actually obsolete. The Englishman William Fox Talbot had already developed

The first daguerreotype camera made for sale. These cameras were made by Alphonse Giroux, a relative of Mme Daguerre. They went on sale on August 19, 1839, and were sold out within a few hours. *Courtesy Kodak Museum, Harrow.*

the negative-positive process that we still use today.

For some time before Daguerre's invention was made public, Talbot had been working on a method of making photographs. When he heard the details of Daguerre's discovery, Talbot rushed to present his own system before the Royal Institution of Great Britain, desiring at least to demonstrate that his method was entirely different from Daguerre's. In an attempt to establish priority, he exhibited a number of photographs he had taken as early as 1835.

Talbot's process was fundamentally the same as Wedgwood's. But where Wedgwood had used silver nitrate only, Talbot coated his paper with a combination of silver nitrate and salt. These two substances combined to form silver chloride, which was much more sensitive to light. This enabled Talbot to succeed where Wedgwood had failed.

Like Daguerre, Talbot discovered that he could fix his photographs by immersing them in a salt solution. And, also like Daguerre, he later substituted sodium thiosulfate. However, when Daguerre announced his process in 1839, Talbot had not yet learned how to develop a latent image. Since his process was a different one, he could not use mercury, as Daguerre did. As a result he was obliged to leave his sensitized paper in the camera until an image became visible.

In 1840, however, Talbot did discover a method of developing the invisible latent images. Like Daguerre he stumbled onto it by accident. Talbot's developing agent, a mixture of gallic acid and silver nitrate, made it possible to reduce exposure times from half an hour or an hour to only a few minutes.

Talbot thought the advance so revolutionary that he gave the process a new name, *calotype*, which he later changed to *talbotype* on the advice of friends.

Since the calotype process produced a negative, it had certain advantages over the daguerreotype. For one, any number of prints could be made from each negative. It was only necessary to oil or wax the paper negative to make it translucent, and to place it on another sheet of sensitized paper. When the two were placed together in the sun, a positive image was formed. Since the negative allowed light to pass through the transparent areas and held back light where the paper was opaque, the tones of the image were reversed a second time. This was certainly simpler than the rephotographing that was necessary if one wished to copy a daguerreotype.

Furthermore, unlike the daguerreotype, the calotype showed no mirror reversal. Scenes were reproduced exactly as they were seen. Calotypes didn't have to be preserved under glass, and they were cheaper and easier to make. In spite of all these advantages, however, the calotype never rivaled the daguerreotype in popularity.

Possibly, the negative-positive process seemed unnecessarily complicated to a world that had just been introduced to photography. The most important consideration, however, was undoubtedly the fact that daguerreotypes were much sharper. People loved to examine them with magnifying glasses in order to pick out details that had gone unnoticed by the naked eye. Since a calotype was printed by allowing light to pass through a paper negative, a certain amount of fuzziness was unavoidable. Furthermore, the texture and imperfections of the paper often showed up in the final print. The

William Henry Fox Talbot, inventor of the negative-positive process in photography. This picture was made by the photographer Moffat of Edinburgh in 1866. *From the Lacock Abbey Collection, The Fox Talbot Museum, Lacock.*

daguerreotype did show a reversal, but this was not thought important; sitters for portraits saw themselves as they did when looking at their reflections in mirror.

Abel Niépce de Saint-Victor, a cousin of Nicéphore Niépce, sealed the doom of the daguerreotype in 1847 when he discovered how to make negatives on glass plates. Three years later a British chemist improved the process. Within a few more years the daguerreotype had been all but replaced. There were further technical improvements after 1850, of course, but by then photography had become essentially the same process we use today.

One further improvement is worth noting: the introduction of roll film by George Eastman in 1888. Eastman marketed the film in a camera he called the Kodak. Each camera came loaded with enough film for a hundred pictures. When all the film had been exposed, the user sent both film and camera back to the factory to have his pictures developed. The negatives and prints were then returned to the photographer, along with the camera, which had been reloaded with another 100-exposure roll of film.

The glass plate has never been entirely supplanted. It is still used frequently in various scientific applications of photography and in holography. Glass plates are especially useful in such fields as spectroscopy and astronomy, where scientists need to be able to make accurate measurements. Because roll film swells when it is immersed in a developer, distances between different points on a negative can be slightly altered. In ordinary picture taking the difference is not noticeable, but when one is trying to determine the wavelength of a spectral line, this discrepancy can be extremely crucial.

Although the basic photographic process has remained essentially the same for more than a hundred years, we may begin to see some changes in the near future. Silver is becoming more scarce and more expensive every year, and eventually it may become necessary to replace the silver process with something cheaper. One suggestion has been something similar to the process used in photocopying machines. Photography based on nitrogen bubbles is another.

But whatever the new system, the scientists and engineers who develop it will spend a lot of money and put in years of research. The silver process has never been supplanted because it provides a simple way of making images of high quality. It will be hard to find something else that can match it.

Six:
The Quantum Theory

There is a story, possibly apocryphal, about the great German theoretical physicist Max Planck. In 1875, when Planck was seventeen and preparing to enter the university, he approached the head of the physics department to discuss his intention to study the subject. The professor, however, gave Planck little encouragement. "Physics is a branch of knowledge that is just about complete," he said. "The important discoveries, all of them, have been made. It is hardly worth entering physics anymore."

The story may or may not be true; it has been told of other renowned physicists and other professors. It is understandable, however, that in 1875 a physics professor might have tried to discourage prospective students.

In the latter part of the nineteenth century, scientists believed that little was left to be discovered. Newton's laws of motion and law of gravitation seemed to provide the last word on those subjects. Scientists understood heat, sound, electricity and magnetism; and Maxwell had successfully explained the nature of light. According to the physicists, it was only necessary now to work out explanations for the few minor remaining puzzles; all of the basic theory had been mapped out.

In spite of the professor's discouraging attitude,

Planck did study physics. Several years after receiving his doctorate, he dutifully set himself the task of trying to clear up one of those little puzzles that was still bedeviling physicists. He soon found that his work was leading him along unforeseen paths. Then in 1900 he announced a discovery that was to revolutionize physics.

Study of the "little puzzle" had led to the unlocking of one of the great secrets of nature. Within a few years all of the smug optimism that had been so common among physicists had vanished. Scientists who had thought they understood virtually everything were suddenly confronted with a host of phenomena about which they understood nothing. By propounding his *quantum theory* in 1900, Planck had unleashed a revolution in physics.

That revolution continues today. Throughout the twentieth century, one discovery has followed another in rapid succession. The atom, thought by nineteenth-century physicists to be indivisible, has turned out to contain a bewildering variety of components. And as yet there is no end in sight; whenever a certain level of understanding has been reached, physicists suddenly find themselves attempting to penetrate another level. The atomic nucleus has been found to contain neutrons, protons and mesons. The neutrons, protons and mesons are believed to be made up of entities called *quarks*. The study of quarks may reveal entities even more fundamental. It is apparent now that the revolution in physics is not likely ever to come to an end, at least not in our time.

It all began near the end of the nineteenth century when Max Planck decided to study a little riddle involving light. It had been known for thousands of years—at the very least—that objects gave off light when heated and that the color changed as the objects became hotter. A piece of iron, for example, becomes first red, then orange, then white. If it is made even hotter, albeit no further changes can be seen with the eye, they can be detected with scientific instruments. If a white-hot piece of iron is heated still further, it will begin to emit invisible ultraviolet radiation in substantial amounts.

The technical term for all this is *black body radiation*. Black body radiation was one of those little puzzles we've been talking about. Scientists understood well enough that when an object became hot, some of its heat energy was transformed into light. But they didn't know how it happened, and they couldn't come up with a mathematical formula that would predict how much light would be radiated at any given wavelength.

The best theory that physicists had been able to devise predicted that a black body would radiate infinite amounts of energy at short wavelengths, that is, in the ultraviolet and beyond. But this was absurd; infinite quantities simply didn't exist. Furthermore, the energy radiated by real black bodies had been observed to decrease at short wavelengths. Obviously the formula didn't work. But no one could discover what was wrong with the theory or why it should predict an "ultraviolet catastrophe."

Physicists had already made numerous attempts to derive other formulas theoretically. So Planck decided to take a different approach. Since there was obviously something wrong with the assumptions the physicists were using, he decided that he would work backward. First, he would use a trial and error procedure to find a formula that worked; then he would try to determine what assumptions had to be made in order to arrive at that formula.

Planck's method worked. The results were so astonishing, however, that even Planck wondered if they could possibly be correct. But he could find nothing wrong with his reasoning, so in 1900 he introduced what was later to be called "the most revolutionary idea that has ever shaken physics"— the quantum theory.

Planck had discovered that light was emitted in little bundles of energy which he called *quanta*. To Planck, who was really a rather conservative scientist, this was very disturbing. It was the kind of result one would expect if light were made up of particles. Yet all the evidence indicated that light was composed of waves.

Unwilling to accept the implications of his own theory, Planck tried to effect a compromise. Although he had demonstrated that light was emitted as quanta, he still assumed that it traveled through space as waves. Planck thought this solution unsatisfactory, but it was the best that he could come up with. He hoped that the mystery would eventually be cleared up and that some way would be found to do away with quanta altogether.

In other words, Planck was a very reluctant revolutionary. In 1905, when Einstein showed that light was indeed made up of particles, that it was emitted as quanta, that it traveled through space as quanta, and that it was absorbed by matter as quanta, Planck refused to believe it. He actually spent years attempting to undo his own work. He never gave up the hope of finding a theoretical method that would lead to the right answer without introducing those unpleasant particles of light.

As time passed, however, there were more and more confirmations of Planck's work. Planck continued to receive recognition for the theory he now detested, and in 1918 he was awarded the Nobel Prize. In his latter years Planck was still trying to combat the trends to which his theory had given support. Near the end of his long life he unwittingly gave a very accurate description of himself when he commented in his *Scientific Autobiography* that "A new scientific truth does not triumph by convincing its opponents and making them see the light, but rather because its opponents eventually die, and a new generation grows up that is familiar with it."

Albert Einstein, who was a little more than twenty years younger than Planck, was a member of that new generation. In 1902 Einstein found a job in the Swiss patent office which turned out not to be especially demanding. He quickly discovered that he could complete a day's work in three or four hours. So naturally he spent the remainder of his time doing theoretical physics. After working half a day on the patent applications, he would turn to his theoretical calculations, scribbling on scraps of paper that could be quickly hidden in a drawer whenever anyone came into the office.

In 1905, after three years in the patent office, Einstein published three papers in the German scientific journal *Annalen der Physik*. Each of the three was to have a profound effect on the course of twentieth-century physics.

One dealt with the theory of Brownian movement (the name given to the random motion of small particles suspended in a fluid). In this paper Einstein conclusively demonstrated the existence of molecules by showing that nothing else could produce the motion; thus, he convinced those members of the scientific community who still had their doubts. In the second paper, Einstein propounded the theory of relativity. The third, which won Einstein the Nobel Prize in 1921, dealt with quanta. In this paper Einstein showed that, contrary to what Planck believed, light really was composed of particles.

The paper on quanta dealt with something called the *photoelectric effect*, another of those little puzzles that nineteenth-century physicists had thought would soon be disposed of. Although it was a fairly simple process—the ejection of electrons from metals exposed to light—no one had been able to come up with a good theoretical explanation. The big stumbling block was that the electrons seemed always to leave any given metal with the same velocity, no matter how weak or how intense the light that produced them. If the light was made brighter, more electrons were produced, but they traveled no faster. There was no way that the wave theory of light could explain this.

Einstein made the simple but revolutionary assumption that the photoelectric effect was caused by the collision of light particles with electrons. He showed that when this assumption was made, theory and experiment agreed perfectly.

This was a bold step. Einstein was making the apparently outlandish suggestion that light could be made up of both waves and particles at the same time. He wasn't attempting to throw out the wave theory; there was too much evidence to support it.

He was saying that although in many situations, light acted as though it were made up of waves, sometimes it behaved as though it were composed of particles instead.

This hardly seemed possible. Scientists had always thought of *wave* and *particle* as two mutually exclusive categories. A grain of sand never showed any wave characteristics; an ocean wave never condensed into solid particles. But now Einstein was telling them that light was somehow made up of both.

In the years that followed, more experiments confirmed Einstein's idea. As the evidence mounted, scientists (except the old diehards like Planck) realized that they had no choice but to accept the paradoxical conclusion that light did have both wave and particle characteristics. Many of them were not especially happy with the situation, and the British physicist Sir William Bragg remarked that there seemed to be no recourse but to believe in waves on Monday, Wednesday and Friday, and to use the particle theory on Tuesday, Thursday and Saturday. Before long some anonymous wit had added, "And on Sunday we pray for enlightenment." With this addition, Bragg's remark became something of a standard joke, one that still appears in modern physics textbooks.

Today scientists no longer speak of quanta of light. The contemporary name for a light particle is *photon* (in analogy with *electron*). The theory of the dual nature of light is still with us, however, although it is no longer considered especially paradoxical.

Wave and *particle* are concepts developed to describe the macroscopic objects we see in the world around us. There is really no reason why they need also be applicable to objects in the microscopic world. In reality light is made of neither waves nor particles, but something which resembles both. Attempts have been made to coin words to name this property (e.g., "wavicle"), but none of them has ever taken hold.

The difficulty in conceiving of light as being simultaneously waves and particles is primarily a semantic one. The nature of light is really very well understood, and once we give up the idea that it has to resemble the things we see in the everyday world, its dual nature is quite easy to accept. The physicists of the early twentieth century may have thought they had to believe in waves on Monday, Wednesday and Friday; but today's scientists see no contradiction in using the combined theory every day of the week.

The investigations of Planck and Einstein were only the beginning of the quantum theory. Although physicists now understood what light was, they still had no idea how it was emitted. That light was given off in some manner by atoms seemed obvious. However, in 1905 no one had any idea what an atom looked like.

The electron had been discovered by the English physicist J. J. Thomson in 1897, but it was still the only known constituent of the atom. The other two fundamental atomic particles, the proton and the neutron, were not to be discovered until 1919 and 1932 respectively. Hence, in 1904 when Thomsom suggested that the atom was composed of a positively charged fluid in which enough negatively charged electrons were embedded to make the atom electrically neutral, it did not seem to be an unreasonable idea. Since Thomson thought of electrons as hard round objects in a relatively soft medium, this is often referred to as the *plum pudding model*.

According to this plum pudding model, electrons (the plums) could vibrate back and forth in the positively charged medium (the pudding). If this was true, then light could be something that was given off when one or another of the electrons slowed down and gave up some of its energy of motion. However, this did not explain why light energy should be emitted in quanta of definite size. It seemed more logical that the electrons would give off energy of varying amounts, depending on how much they were slowed.

More information was obviously needed. However, it wasn't so easy to study the atom. One couldn't put an atom under the microscope and look at it. Obviously, some indirect method had to be found.

Just such a method was discovered by Ernest Rutherford, a physicist from New Zealand who was then director of the physics laboratory at the University of Manchester. Rutherford reasoned that if you want to examine something you can't see, you should poke it. When it has been prodded enough, you should have a good idea of its size and shape.

Rutherford prodded the atom by directing a beam of alpha particles (at the time, neither Rutherford nor anyone else knew what these were, hence the noncommittal designation "alpha"; later, they turned out to be helium nuclei) onto a sheet of gold foil. Those alpha particles which were deflected by the atoms in the foil could be observed by placing a fluorescent screen in their path. Whenever an alpha

Ernest Rutherford, the British physicist who discovered the nucleus of the atom. *Courtesy Niels Bohr Library, American Institute of Physics.*

particle hit the screen, it produced a tiny scintillation of light that could be observed through a low-power microscope.

This method has been likened to shooting bullets

at a haystack in which a metal object has been hidden. Most of the bullets will pass right through, but those which strike the object will bounce off in various directions. If enough bullets are fired, it should be possible to learn something of the object's size, shape and location.

Rutherford anticipated that the atom would be more or less as Thomson had described it. And if it was, his alpha particles would plow right through the "pudding." Some of them might be deflected when they encountered electrons, but since alpha particles were thousands of times heavier than electrons, they would not be deflected very much. In a way this is analogous to a collision between a battleship and a small sailboat. No matter how many of the latter the battleship runs into, it is not going to be thrown very much off course.

Rutherford and his students worked on the experiment for years. It was tedious work. In order to obtain accurate results, they had to observe the deflections of thousands of alpha particles. And since Rutherford possessed none of today's sophisticated electronic detection equipment, it all had to be done by eye. The experimenter was obliged to sit in a dark room for hours, patiently waiting for the scintillations to appear on the screen and keeping an accurate count of them.

The work required a great deal more patience than Rutherford possessed. After counting scintillations himself a few times, he left the task to his students. He was, however, an active participant; the idea for this piece of research had been his, and it was he who did the mathematical calculations that were necessary to interpret the results. If he preferred to let his students perform the more menial tasks, one can't blame him. Professors commonly do the same thing today.

As expected, most of the alpha particles passed through the foil with little or no deflection. But, surprisingly, one in every several thousand veered off at a wide angle, and some actually bounced back in the direction from which they had come. Rutherford realized that there was no way that the Thomson model of the atom could explain this result. He commented later that it was "almost as incredible as if you fired a fifteen-inch shell at a piece of tissue paper and it came back and hit you."

Rutherford's reasoning told him that this could mean only one thing: the alpha particles had encountered a tremendously strong electric field. This could happen only if the positive charge in an atom was concentrated in a small nucleus. The Thomson model had to be wrong; the electrons were not embedded in a pudding. They circled a small nucleus in a manner analogous to planets revolving around the sun.

In 1911 Rutherford announced his discovery of the nucleus. However, this still did nothing to explain how an atom emitted light. In fact, the planetary model of the atom that grew out of Rutherford's work seemed to make the problem even more puzzling.

According to Maxwell's electromagnetic theory, a negatively charged electron circling a positive nucleus should emit light continuously. As it did so, it would lose energy and spiral in toward the nucleus. A simple calculation seemed to indicate that an electron should remain in orbit for only a very small fraction of a second. If Rutherford's model was correct and the electromagnetic theory was true, then atoms shouldn't exist. If any did happen to come into existence, they would be destroyed as soon as they were created. And for whatever period of time they did exist, all matter would continuously glow with light.

To the scientists of the day it must have seemed that physics was rapidly passing from one crisis to another. A little more than a decade earlier it had seemed that nearly everything had been explained. But now they were being asked to believe not only that light had a dual character, but also that the atom had apparently contradictory properties. Many of them certainly must have thrown up their hands and wished they could do away with atoms and quanta altogether. Unfortunately, they couldn't; the evidence for the existence of atoms and quanta was too great.

Quite a few physicists did, however, distrust Rutherford's model of the atom. In their view, to ask "What is the atom?" was to ask a question that would never be answered. The contradictions that Rutherford's research had introduced indicated to them that inquiries into the ultimate nature of matter were profitless. Confronted with things that they couldn't understand, they desired to pursue such subjects no further.

Then in 1913 the Danish physicist Niels Bohr, one of Rutherford's former students, removed all the difficulties and made atomic research the primary concern of physics. In that year Bohr proposed his quantum theory of the atom.

Bohr got rid of the problems that had been plaguing atomic physics by the simple expedient of legislating them away. He made the audacious assumption that an electron in a circular orbit would

Niels Bohr (left) and Max Planck, two of the founders of the quantum theory. *Courtesy Niels Bohr Library, Margrethe Bohr Collection.*

not radiate light while it was part of an atom. At all other times it would obey Maxwell's theory, but once it started to circle a nucleus, that theory no longer applied.

Bohr could give no reason why this should be so. He pointed out, however, that such an assumption had to be made; there was no other way out.

In propounding his theory, Bohr assumed further that only certain electron orbits were possible. The electron could revolve around the nucleus only at rigidly defined distances that were given by a simple mathematical formula; it could not be located anywhere in between. In other words, the atom was *quantized*.

An electron, Bohr went on, could jump from one orbit to another, emitting or absorbing a quantum of light in the process. This happened instantaneously; when the jump was made, the electron disappeared and rematerialized somewhere else. Bohr's theory implied that it was senseless to ask how the electron got from one orbit to another. One only knew that it did.

It sounded like pure fantasy. Nevertheless, the theory had to be taken seriously because, when worked out in detail, it explained how quanta of light were emitted. Furthermore, it accurately predicted just what wavelengths of light an atom could emit. It might seem bizarre, but it worked.

For decades spectroscopists had been studying the light given off by atoms. Now Bohr's theory seemed, for the first time, to give some sort of structure to their findings. What had been a bewildering mass of experimental data could at last be interpreted by applying the new atomic theory.

Before the advent of Bohr's quantum theory, no one had been able to say why an atom should emit certain wavelengths and not others. By solving the problem, Bohr founded the field that was to become atomic physics. He later received the Nobel Prize for this achievement, joining that select group which included Planck, Einstein, and Rutherford.

No theory in physics has ever proved capable of giving any final answers. However much has been learned, it is always possible to penetrate more deeply into the secrets of nature. And so in time Bohr's theory was replaced by one that was even more sophisticated and which interpreted experimental data even more accurately. This theory, developed by the German physicists Werner Heisenberg and Erwin Schrödinger in the mid-1920s, is called *quantum mechanics*. The word *mechanics* distinguishes it from Bohr's quantum *theory*.

In quantum mechanics the ideas of quantized atoms and electron jumps are retained. As in Bohr's theory, light is assumed to be emitted when an electron goes from one orbit to another. The major difference is that in quantum mechanics the electron is pictured not as a particle but as a packet of waves.

Virtually all of modern physics is based on quantum mechanics. The theory not only explains the emission of light and the behavior of atoms; it has also been successfully applied in such diverse areas as the superfluidity of liquid helium and the processes that take place in the interiors of stars. Solid state physics, which was responsible for the development of the transistor, is based on quantum mechanics. So are theories of such subnuclear particles as protons, neutrons and mesons, and their hypothetical constituents, the quark. It is quite accurate to say, therefore, that modern physics began with Max Planck's study of the emission of light.

Although the twentieth-century revolution in physics is a subject of enormous interest, we are primarily concerned with discoveries concerning the nature of light. It might not be a bad idea, therefore, to summarize what has been said about light in this chapter. In the process we will fill in a few additional details.

Light is emitted by atoms in the form of quanta, or photons, as they are called today. Light, however, is neither a particle nor a wave; it has characteristics common to both.

According to Bohr's theory, light is emitted by an atom when an electron jumps from one orbit to another. The orbit closest to the nucleus, called the *ground state*, has the lowest energy. When an electron absorbs energy in one of its various forms—heat, light or electrical energy—it is kicked into a higher orbit. In an extremely short time, approximately a hundred-millionth of a second, the electron will jump back down to the ground state, giving off a photon of light as it does. The energy of the photon is described by a simple formula, originally discovered by Planck:

$$E = hf$$

where E is the energy, f is the frequency of the light, and h is a small number called *Planck's constant*. Since the frequency of a wave equals the speed of light divided by the wavelength, photons of a given wavelength always have the same energy. The smaller the wavelength, the greater this energy. This explains why very short wavelength waves, such as those of X rays or gamma rays, have such great energy, and why radio waves, which have very long wavelengths, must be fed into amplifiers if they are to be detected at all.

Planck's formula is interesting also because it shows that even when one tries to think of light as being made up of photons, it is impossible to avoid speaking of frequency, which is a wave characteristic. The wave and particle pictures of light seem to be inseparable.

When quantum mechanics replaced Bohr's theory, the explanation of the emission of light remained fundamentally the same. The only real difference is one of terminology. Since the electron is now thought of as a wave rather than as a particle revolving around a nucleus, one normally uses such terms as *state* and *orbital* in place of *orbit*. The emitted photons have an energy equal to hf, just as they

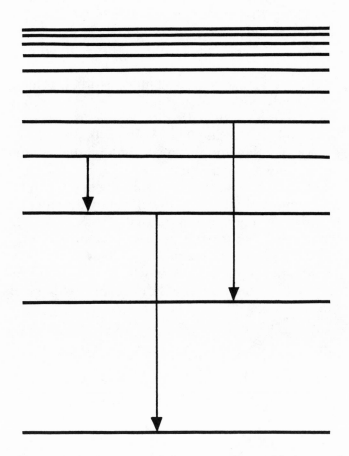

do in Bohr's theory, and there is still a stable ground state of lowest energy. States of higher energy are called *excited states*. The possible energy levels of an atom are often represented by diagrams like the preceding. The arrows represent jumps, or transitions, in which light of various frequencies is given off.

There is still, it seems, one question we haven't answered. If both light and matter are composed of something that is neither waves nor particles but which resembles both, then why do light and matter seem so different? Why is it not possible to convert light into matter, and matter into light?

The answer, as anyone familiar with Einstein's equation $E = mc^2$ should realize, is that they are not so different, and that it *is* possible to transform light and matter into each other. In his special theory of relativity, Einstein showed that matter and energy are equivalent. Since light is a form of energy, if we are willing to forsake the mathematical precision of physics for the comparative vagueness of ordinary language, it is possible to say that "light (or energy) and matter are, in some sense, only different forms of the same thing." This topic will be fully discussed in Chapter 9, which deals with Einstein's two theories of relativity and with black holes.

Seven:
Lasers

In the late 1950s a small number of scientists were conducting research with a recently invented device called the *maser* which was used to produce microwaves. The letters that made up the word stood for "Microwave Amplification by Stimulated Emission of Radiation." At least that was the official version; according to a joke current at the time, they really stood for "Money Acquisition Scheme for Expensive Research."

This might have been an attempt at scientific humor, or possibly it was a nervous admission on the part of the maser researchers that they hadn't come up with much to justify all the money they had spent.

If the latter was the case, they shouldn't have felt bad about it. Even though the maser is still not very important, the money was well spent. In 1960 American and Russian scientists, working independently, used the principles on which the maser was based to develop a device that produced light rather than microwaves. By analogy this was called the *laser*, for "Light Amplification by Stimulated Emission of Radiation."

Although the maser had only limited application, the opposite quickly proved to be the case with the laser. By the early 1960s it had become apparent that the laser would turn out to be one of the most significant inventions of the twentieth century. Since then it has more than fulfilled its promise. Numerous uses have been found for lasers in scientific research. They are used in industry for cutting cloth and welding steel, and surgeons use them for delicate operations. They ring up prices in supermarkets and transmit telephone messages. They are being made into futuristic "death-ray" weapons by the military. In the coming years they may be used to produce fusion power, to propel space vehicles, and to facilitate interstellar communication. Lasers have even been adapted to such trivial but hitherto insoluble problems as the eradication of tattoos.

Yet a laser is a very simple device. It does nothing more than emit a kind of pure light by a process called *stimulated emission*. It would not be inaccurate to say that it is its very simplicity which gives laser light its unique properties. Most light sources radiate in a relatively complicated manner. As a result their light is not nearly so pure as that produced by a laser.

Stimulated emission can easily be described by quantum mechanics. We need only recall that when an atom passes from a higher energy state to a lower one, it gives off a photon of light. If this photon strikes a second atom that is in the same original excited state, the photon will stimulate emission from the second atom, producing a photon exactly like the first. As these photons strike additional atoms, we will have three photons, then four, then five, and finally billions.

One might expect such a process to take place all the time; after all, atoms go into excited states whenever they absorb even a little energy. All it takes is some light or a little heat or the energy from a chemical reaction. Why then is not stimulated emission a more frequently encountered process? Why isn't every flowerpot a laser? and every cocktail glass? and every basketball?

The answer to these questions is that all those things would act as lasers if it were not for the fact that atoms normally remain in excited states for extremely short periods of time. Atoms generally will not wait for photons to come along and stimulate them. Instead they give up their energy about a hundred-millionth of a second after they absorb it. Before any photons can collide with an atom, it will have given off its own photon spontaneously.

Some atoms, however, have energy states that are called *metastable*. If an atom is kicked into one of these states, it will stay there for a thousandth or a hundredth or a tenth of a second—a very long time by atomic standards.

Suppose we somehow manage to put a whole collection of atoms into such metastable states. Suppose further that they stay there for an average of a thousandth of a second. Some of the atoms will emit photons much sooner than others do. With the first emission, laser action begins. The photon from the first atom stimulates another, which stimulates another, which stimulates another, and so on. Within a very short time, all of the atoms will have given off their light.

All of the photons emitted by a laser will be exactly alike. If we picture this emitted light as waves rather than particles, we can understand that all of the light will be of a single, very narrowly defined wavelength. Furthermore, all of the light waves will be in phase with one another; troughs will line up with troughs, and crests with crests.

Two little pieces of scientific jargon are frequently used to describe these properties: *monochromatic* and *coherent*. *Monochromatic* means "of a single wavelength," while *coherent* is used to describe light in which all of the individual waves are in phase.

Laser light bears little resemblance to that which is emitted by ordinary sources—incandescent or fluorescent lamps, for example. The light given off by such lamps is a jumble of different wavelengths, and even the light that is of the same wavelength is rarely in phase. Sometimes the crest of one wave will be superimposed on the trough of another, and the two will cancel each other out. Sometimes the opposite will be the case. If a laser is compared to a well-trained choir whose members sing in unison, then an ordinary light source resembles a group of singers that allows its members to sing different tunes at whatever tempo they wish.

Laser light has yet another property that is important for some applications, though it is not related to the process of stimulated emission in any fundamental way. Laser light is emitted in bundles of parallel rays. As a result a laser beam will not spread out to any appreciable extent. It can travel for miles and still retain its original narrow width.

The reason why lasers emit parallel light can perhaps best be explained by means of an example. The one that will be described is the ruby laser; all other lasers are constructed in similar ways.

The first lasers were made from cylindrical crystals of synthetic ruby. Two mirrors, one of them half-silvered, were placed at the ends of the cylinder, so that the light would bounce back and forth within the ruby many times before it made its way out through the half-silvered mirror. This caused a buildup in the intensity of the light. A ray of light that goes back and forth through the ruby a number of times can stimulate emission in a greater number of atoms than can one that passes through only once. This obviously also has the effect of making the light parallel. The light that escapes is light that has been bouncing back and forth along the axis of the ruby. A small amount is emitted in oblique directions; it is less intense because it escapes from the rod before it can collide with an appreciable number of atoms.

In order to make these rubies *lase*, it was only necessary to illuminate them with light from flash tubes similar to those contained in the electronic flash units used in photography. This light pushed the atoms into metastable states. After a fraction of a second, laser action took place spontaneously.

A ruby laser does not emit light continuously; it gives off short pulses that last for times as short as a millionth of a second. Sometimes this is desirable; however, there are some applications for which it is necessary to have an uninterrupted beam of light. This became possible in 1961 when the first continuous gas laser, a mixture of helium and neon in a narrow tube, was made to produce coherent light.

Since that time laser action has been produced in hundreds of different substances. There are crystal lasers like the ruby, and there are gas lasers, liquid lasers, semiconductor lasers and glass lasers. Some are powered by flash tubes; others derive their energy from electrical discharges. There are lasers which produce light of relatively low intensity, and there are lasers powerful enough to cut steel.

Today lasers are a multimillion-dollar industry. Yet when the laser was invented, no one could think of a practical use for it: it was a solution for which there was no problem. As a result the first applications of the laser were in the area of scientific research.

Since the laser could produce light of a very precise wavelength, it could make measurements of a greater accuracy than had previously been thought possible. One of the most dramatic experiments took place in 1973, when lasers were used to measure the distance from the earth to the moon to an accuracy of six inches.

The astronauts who had landed on the moon during the Apollo 11, 14 and 15 flights placed reflectors at various points on the moon's surface. These were illuminated with pulses of laser light that were directed at the moon through telescopes on earth. When the reflected laser pulses were received on earth, the elapsed time was measured and the distance to the moon computed. Since this 235,000-mile distance varies, the experiment was performed repeatedly in order to get an accurate picture of the moon's motion.

Such an experiment would not have been possible with any existing light source other than a laser, because the beam would spread out too much, and not enough light would be reflected back to be detectable. Since a laser emits parallel rays, however, a beam originally a few millimeters wide had spread out to a diameter of only two miles by the time it reached the moon. Only about a billionth of the light that was sent out managed to make it back to the detectors on earth, but this was enough to be measurable.

The accuracy of the measurements had been increased by the findings of a group of scientists working at the National Bureau of Standards in Boulder, Colorado, in 1971, who had used lasers to measure the speed of light with undreamed-of precision. The Bureau of Standards group developed a way to stabilize the already precise wavelengths of laser light even further, thus obtaining a figure a hundred times more accurate than the older accepted value, 299,792.5 kilometers per second. The new figure for the velocity of light was 299,792.4562 kilometers or 186,282.3960 miles per second.

Someday lasers may be used to communicate with intelligent life elsewhere in the universe. Although most scientists believe that there are other advanced civilizations in the galaxy and that at least some of them must be sending messages across interstellar distances, no such signals have been detected. There is every reason to think, however, that they eventually will be. If this should happen, the benefits to humanity would be incalculable.

Most of the work done in interstellar communication so far has involved looking for radio signals. Unfortunately, a radio telescope can be tuned to only one frequency at a time, and it would take years to examine all the frequencies that might be used. Moreover, only a small area of the sky can be examined at any one time. If scientists were to look at all the possible frequencies and search all sections of the sky, expenditures of billions of dollars would be required, and the task might take ten years or more. Since no government has proved willing to appropriate the necessary money, the scientists can listen for messages from only a small number of stars. As a result, some of them have begun looking for flashes of laser light coming to us from the depths of space.

Although radio signals are easier to transmit, laser messages would be quite a bit simpler to detect. It is possible that our extraterrestrial friends realize that governments are reluctant to appropriate large sums of money when they are not sure there will be any results. Whatever kind of creatures they are, it may be that they have had similar difficulties on their own planet; hence, they may want to make things easy for us by sending laser signals.

All this is speculation, of course. We have no idea what these beings are like or what kind of psychology they may use. For all we know, the only other advanced civilizations in this galaxy may be made up of intelligent plants who have no desire whatsoever to talk to anyone else. Many scientists believe, however, that it is worth taking a look to see if the laser signals are there.

One might think that a laser message from space would be impossible to detect, that light from another planet would be obscured by the light of the sun around which the planet revolved. This would indeed be the case if any light source other than a laser were used. However, since a laser emits a single wavelength, it should not be too hard to produce a beam whose brightness *at that wavelength* is greater than that of a sun. Sunlight, after all, is spread out over many wavelengths, while laser light is extremely concentrated.

There are other factors which make detection of laser light relatively easy. One is the "static" caused by interstellar hydrogen which interferes with the reception of radio messages in the channels most likely to be used. Light signals, by comparison, are static-free.

But of course there's a catch. The most logical wavelength for interstellar communication lies in the ultraviolet region of the spectrum. And ultraviolet radiation is absorbed by the earth's atmosphere. Thus, any ultraviolet laser light that did reach the earth's surface would be too weak to detect. In order to look for laser messages, it is necessary to place detectors in orbiting satellites.

In 1975 an ultraviolet telescope from Princeton University was placed on Copernicus, the Orbiting Astronomical Observatory satellite. However, it was able to search for messages from only three stars, each about eleven light-years from earth. Since it may be necessary to look at millions—or billions—of stars before any messages are detected, the possibility of entering into conversations with other life forms will remain slim for some time to come. Although detecting laser signals might very well be easier and cheaper than searching all possible radio frequencies, at present it isn't easy or cheap enough. In all likelihood we will have to wait until we have permanent astronomical observatories in space before such a project has any chance of success.

There are numerous applications of lasers in scientific fields other than physics. "Tunable" lasers which can be set to any of a number of different wavelengths of light have proved to be very useful to chemists. A laser beam set at a particular wavelength and directed into a vat of chemicals can cause one kind of atom to react while all the others remain inert. Biologists are focusing laser light down to an intense spot that is only about 0.0005 millimeter—about one wavelength of light—in di-

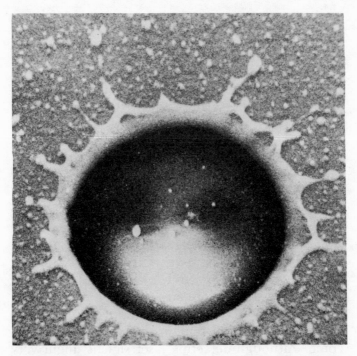

Since lasers produce beams of parallel light, they can be focused to tiny points. This hole, about one-tenth the diameter of a human hair, was produced by a single laser pulse directed on a coating of bismuth metal over a plastic substrate. *Courtesy Bell Laboratories.*

ameter. This makes it possible for them to use lasers as tiny scalpels for surgery on chromosomes or on living cells.

Although many of the scientific applications of laser light are exciting, the possibility of making lasers into "death rays" has received the greatest publicity. Such stories are more than mere speculation. Although such projects are naturally shrouded in secrecy, it seems certain that the military has made quite a bit of progress toward the development of death-ray weapons. Other kinds of laser weapons have already been used. "Smart bombs" guided by laser light were part of the military arsenal during the closing days of the Vietnam war.

Speculation about death rays began in the early sixties when it was discovered that lasers could be used to burn holes in objects such as razor blades. These early lasers, however, were inefficient devices of very little power. If they were able to make holes in Super Blue Blades, it was because laser beams could be focused down to very tiny spots, not because they produced light of any great intensity.

More powerful lasers were needed before death rays could become more than a remote possibility. No one has ever disabled a tank by burning pinholes in its armor. Such weapons would be equally useless against an airplane or a missile or even an enemy soldier. What was needed was a laser that could produce damage over a larger area, say the size of a basketball.

A step forward (or backward, depending on your feelings about military applications of scientific discoveries) was made in 1964 when the powerful carbon dioxide laser was developed. The CO_2 laser was capable of producing a beam of far greater intensity than any that had been created before. We can be sure that the military was aware of the uses to which the CO_2 laser could be put, but at that point the work was undoubtedly classified "Top Secret."

Beginning in the early 1960s, the military had a weapons development project, code-named "Eighth Card," headquartered at Kirtland Air Force Base near Albuquerque, New Mexico. Although the project was administered by the Air Force, all of the

U.S. armed forces contributed to it under the direction of the Defense Department's Advanced Research Projects Agency.

In 1972 the trade publication *Laser Focus* reported that scientists working on the project were using lasers to set fire to wooden planks two miles away, and that the accuracy of the weapon being tested was so great that it could hit an object the size of a playing card as it waved at the end of a twenty-foot pole one mile away. Later there were reports that the laser was firing successfully at unmanned aircraft.

Since 1972, work on laser weapons has continued, but only those who have access to top-secret documents are in a position to know much about them. We can be sure, however, that sooner or later some kind of full-scale development of these weapons will be proposed. When that happens, it is likely that there will be a furor much like the one that surrounded the proposed deployment of the antiballistic missile (ABM) system. The danger lies in the possibility that the refinement of laser weapons would accelerate the arms race and perhaps imperil the SALT agreements between the U.S. and the Soviet Union.

Although the Pentagon isn't saying anything about the weapons it plans to develop, it is possible to make some guesses as to their nature. In all probability, laser weapons would be intended for use in space to shoot down satellites or to intercept ballistic missiles.

It would be possible to use laser weapons on terrestrial battlefields, of course, but certain limitations might make them less effective than more conventional weapons. Lasers will not readily pene-

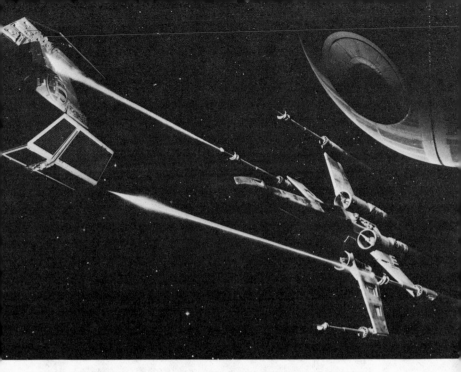

Although *Star Wars* was science fiction, laser weapons are not. Such weapons could be adapted for use in space in the very near future. The beams they produced would be invisible, however, since there are no dust particles in space to scatter the light. *Courtesy Twentieth Century–Fox. Copyright © 1977 Star Wars Twentieth Century–Fox Film Corp. All rights reserved.*

trate clouds or even thin fog, for example, and even on clear days laser beams are subject to dispersion in the atmosphere and can be deflected by mirrors or shiny surfaces. It is very likely, therefore, that for the next few years at least the use of lasers in earthbound weapons systems will be limited to range-finding and homing devices such as that used in the "smart bombs" mentioned previously.

The situation might change rapidly, however, if it proved possible to develop gamma-ray or X-ray lasers. Admittedly, this is very speculative, but these lasers, unlike those that produce visible light, would not be subject to atmospheric dispersion. Since they would emit very high energy radiation, they could be made into weapons that nothing could withstand.

Such weapons might come into existence sooner than we think. According to the journal *Science News*, Soviet scientists have been conducting research in this area for several years now. If that is true, we can be sure that the U.S. military is aware of what they are doing and that it is not allowing itself to fall very far behind.

I should probably mention that while I was a graduate student at the University of New Mexico, I spent the summer of 1963 working with the Air Force laser weapon project at Kirtland Air Force Base. It would be impossible for me to divulge anything about the work that was being done, however. The people who work on such projects don't talk much about the progress of their research, certainly not to graduate students with low-level security clearances. I was given a relatively unimportant research problem of my own to work on, and I did not come into contact with any other work.

In fact, I had not even heard the code name Eighth Card until I began doing the research for this chapter, some fifteen years after I had worked on the project. I did learn one thing worth mentioning, however. I began to understand how fatally easy it is for a scientist to become immersed in the purely scientific aspects of the work he is doing and thus to become oblivious to its future military applications. If this were any reasonable kind of world, we would all find the prospect of doing military research repugnant, but, unfortunately, it is much too easy to let the challenge of solving the problem obscure the human ramifications of the solution.

Of course, high-energy lasers have numerous nonmilitary applications. They are widely used in industry for welding and cutting steel. Lasers can make deeper welds than the more conventional devices, and they produce better bonding. When used to cut metals, they cut more cleanly. The Ford Motor Company, just one of many manufacturers using lasers, welds Ford and Mercury underbodies with laser beams. Its major competitor, General Motors, has made use of a CO_2 laser for cutting auto parts. Watchmakers use lasers to cut holes in jewel bearings. Similar lasers drill holes in the diamond dies used to manufacture wire. The garment industry uses lasers to cut clothing. A laser system is even used to drill holes in nipples for baby bottles. Not only is the process faster than those used previously, it is also more sanitary because it leaves no residue.

The lower power lasers also have numerous uses. They can be used to align sewers, to guide dredging barges, and to inspect buttons for defects as they pass by on assembly lines. The scanning devices now used in many supermarkets to read barred product codes make use of lasers. And by the end of the century lasers will probably be transmitting most of our telephone messages.

Certainly, most people are aware that the telephone was invented by Alexander Graham Bell just over a hundred years ago, but relatively few know that Bell also invented a contraption called the photophone which could transmit human speech on a beam of light. Although the photophone

worked well enough, Bell eventually gave up on it because the transmission of electrical impulses along copper wires seemed much more practical.

The development of the laser evoked renewed interest in the idea of transmitting messages by means of beams of light. Before long this method will be much cheaper than electrical transmission. Right now the best telephone cables can carry something like a hundred thousand two-way conversations. But cables of glass fibers carrying laser beams may prove to have capacities of up to a hundred million conversations. Furthermore, glass is relatively cheap and light compared to metal, and it is immune to electrical interference. Although there are still problems to be solved, they do not seem to be serious ones. There is every indication that the present methods of transmitting telephone messages will soon become things of the past.

One of the most widely publicized applications of the laser is holography, a method of creating three-dimensional pictures without using either camera or lens. Contrary to popular belief, holography is not a new process. Its basic principles were described in a paper published in 1948 in the British scientific journal *Nature*. At the time there were no light sources sufficiently coherent to make the process workable. With the invention of the laser, however, holography immediately became practical.

To make a hologram, it is only necessary to ex-

Lasers can be used to machine miniaturized electronic circuits. The rooster-tail pattern in this photograph is excess material vaporized by a laser beam during a laser-scribing operation. *Courtesy Bell Laboratories.*

pose a photographic plate to light reflected from an object that has been illuminated with a laser. Part of the light from the laser, called the *reference beam*, is split off so that it bypasses the object and reaches the plate directly. The reflected light and the reference beam combine to produce an interference pattern on the plate which, although it looks not at all like a picture, contains enough information to create a three-dimensional image.

When the plate is developed and examined in ordinary light, it seems to contain nothing but a number of wavy lines. But if a laser like the first is made to shine through the plate, called a *hologram*, a three-dimensional image is created.

Holography differs from ordinary photography in that it does not make use of lenses to form images. The photographic plate records the wave front of light that spreads out from the object. When the hologram is illuminated by the second laser, this wave front is reconstructed. It would not be too inaccurate to say that the hologram "traps" light from the object, and that this trapped light is released by the second laser when the picture is viewed.

Holograms have one very surprising property. If the plate is broken into a number of small pieces, the original image can be recreated from each individual piece, no matter how small. To be sure, there is a certain loss in resolution.

Holography may eventually give us three-dimensional color television, although such a development almost certainly lies decades in the future, because some very difficult technical problems will need to be solved. Holography has, however, found a number of applications. For example, it can be used to detect defective automobile tires, to encode and store computer data, and to study stress patterns in steel and other structural materials.

The laser has numerous applications in the field of medicine. Surgeons use lasers as scalpels for performing such difficult operations as the removal of growths on vocal cords. Since a laser cauterizes blood vessels as it cuts through tissue, bleeding during an operation can be eliminated. Bloodless surgery is especially important for operations on the liver, which have always been very dangerous because the liver contains so many blood vessels.

When a patient is operated on for a bleeding ulcer, surgical opening of the stomach sometimes causes more damage than the ulcer itself. Now there is hope that this can be avoided by using lasers. Soon, medical researchers say, a surgeon will have only to slip a tube similar to those used for the transmission of phone messages down the patient's throat into his stomach. After the ulcer has been located, bursts of laser light can be transmitted down the tube to cauterize the bleeding tissue.

The laser also seems to have great potential for the treatment of cancer. Unlike X rays or gamma rays, a laser beam can act selectively on body tissue, destroying cancer cells without doing excessive damage to the healthy cells which surround them. Lasers can be used to irradiate tumors or as scalpels for cutting them out. They are already being used to burn off skin cancers, which, because they are frequently darker than the surrounding skin, absorb the laser radiation selectively.

Lasers can be used in the treatment of detached retinas. When laser bursts are directed into the eye, they come to a focus on the retina and "spot-weld"

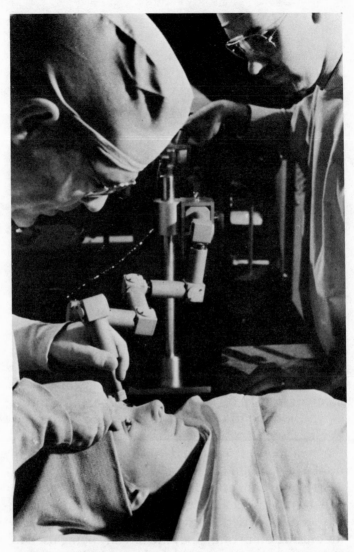

Ophthalmologist Dr. Francis L'Esperance, assisted by Dr. Gordon Kelley, prepares to operate on an eye tumor at New York's Columbia-Presbyterian Medical Center. *Courtesy Bell Laboratories.*

it back into place. The laser can also be used in the treatment of diabetic retinopathy, a disease which often causes blindness in diabetics. Light from an argon laser is used to coagulate vision-obscuring blood vessels.

Work has been done on developing a kind of laser radar for blind people. A laser device has been developed which emits invisible infrared light and is small enough to fit on eyeglass frames. When the laser light strikes an obstacle, it is reflected back to a receiver that triggers a vibrating pin against the blind person's finger, or sounds a tone in his ear. It remains to be seen, however, whether laser systems of this type will replace the white cane. The latter seems to have the advantage of being quite a bit simpler and cheaper.

The Air Force is developing a sun-powered laser for space communication. It should be able to transmit data of any kind, including voice messages and television signals. Sun-powered lasers may also be used one day for the propulsion of space vehicles. If this proves feasible, the exploration of space will be much easier and cheaper. Such devices allow separation of the energy source from the vehicle, making possible lower payloads. The lasers could beam energy to the ship from aircraft or space stations.

Such devices probably would not be used to propel earth-launched space vehicles, however. The energy requirements would be too high, and the lasers would face the same problems with atmospheric dispersion as have been encountered in weapons research. Similar lasers could be used to transmit power over long distances, but again it is more likely that they would be designed primarily for use in space.

Lasers may eventually make a significant contribution to the solution of our energy problems by making controlled fusion possible. Laser-induced fusion has, in fact, already been achieved at the University of California's Lawrence Livermore Laboratory. Although it may be decades before a commercially feasible process is developed, progress in that direction is being made.

There are two types of nuclear energy, *fission* and *fusion*. Fission, a process which involves the splitting of uranium or plutonium atoms, is used in atomic bombs and in nuclear reactors. Fusion, the process which takes place in hydrogen bombs, is also the mechanism by which the sun produces its heat and light. It involves the fusing of deuterium or tritium, two isotopes of heavy hydrogen, to make helium. Fusion not only produces more energy than fission, it also uses a fuel which we possess in virtually limitless amounts. Deuterium and tritium can be obtained from ordinary water. Furthermore, since the end product, helium, is not radioactive, some of the problems with radioactive waste that are associated with fission reactors might be avoided.

Fission reactors are now fairly common. However, scientists have been struggling to create controlled fusion for years without much success until just recently. In the past two decades an enormous amount of work has gone into the search for ways to confine fusion reactions in strong magnetic fields. Although such attempts might eventually be successful, far better results in much less time have been obtained with laser fusion.

At the Lawrence Livermore Laboratory powerful lasers are focused on tiny glass of plastic pellets containing deuterium and tritium fuel. When a laser

Lawrence Livermore Laboratory's "Shiva" laser complex, which is being used in experiments on the production of controlled fusion. *Courtesy Lawrence Livermore Laboratory, University of California, and the U.S. Department of Energy.*

burst strikes its target, the material that makes up the surface of the pellet explodes. The explosion acts on the deuterium and tritium with an inward force, compressing it to something like one ten-thousandth of its original volume. The compression heats the fuel to temperatures in the neighborhood of 100,000,000° C, and fusion is induced.

The pellets are relatively cheap and easy to make; however, the lasers which produce the beams that strike the pellets are not. These lasers are huge devices capable of producing trillions of watts of energy in short pulses lasting only about a billionth of a second each. Even so, the fusion reactions obtained so far have not reached the break-even point; more power is required to drive the lasers than is obtained from the fusion. Thus, commercially feasible fusion reactors will not be available before the 1990s, possibly not until the twenty-first century.

While technological advances solve a host of problems, they always seem to create new problems of their own. Fusion power has been touted as a safe, cheap, clean source of energy. However, it could very well turn out to have an environmental impact as great as that of present power production methods.

We would do well to recall that when nuclear fission reactors were first introduced, they too were supposed to provide safe, clean power. They were compared favorably with oil and coal, and it was pointed out that they produced "no pollution." Within a very short time, though, it was recognized that the problems of radioactive waste disposal and thermal pollution associated with the reactors were just as difficult to deal with as the kinds of pollution associated with fossil fuels. As a result nuclear power has never fulfilled its initial promise.

Fusion reactors produce heat, also. Hence, thermal pollution problems might be as great as with fission reactors. And although it is true that helium, the end product of the fusion reaction, is not radioactive, there would still be radioactive waste to be disposed of. The fusion reaction produces a large number of high-energy neutrons that would irradiate any chamber in which it was contained.

There is, however, some hope that such problems could be solved, for methods of making nuclear energy environmentally safer are in the works. It should come as no surprise that lasers are involved here also.

At present, radioactive wastes are subjected to complicated chemical treatment designed to separate long-lived from short-lived isotopes. Not only is the process inefficient—some forty steps are required—it also produces a greater volume of radioactive material than was there at the beginning. Because chemicals have to be added to the waste, the amount of radioactive matter grows greater at every step.

A simple laser process promises to make "cleaner" isotope separation possible. The isotopes could be vaporized in specially built ovens, and then lasers would selectively ionize the atoms of only one isotope. Once ionized, these atoms could be easily separated from the rest. The process could be repeated with other kinds of atoms as many times as necessary.

We could have discussed numerous other applications of lasers: for example, their use in determining time of death in a corpse; or in removing the deposits that industrial pollutants have formed on Renaissance sculptures. A list of applications would be a very long one; even as this chapter was being written, the list may have been growing. The examples given, however, should be sufficient to

indicate that the laser is likely to affect our lives—
for better or for worse—in numerous significant
ways in the years to come.

It would be useless to speculate to any great degree as to the precise nature of these effects. Predictions about the technology of the future are
notoriously subject to error. Too often they manage
to be simultaneously overly optimistic and overly
conservative, making wild projections in some areas
while failing to foresee significant advances in
others.

All in all, it is probably just as well to leave the job
of prediction to the psychics and the astrologers.
There is, however, one prediction that can be made
with assurance: lasers will be very much with us in
the years to come, for they have begun to propagate
themselves; they are now being used in the con-
struction of new lasers.

Eight:
Light from the Creation

In 1917 Albert Einstein, who had recently completed work on his general theory of relativity, was trying to formulate equations that would describe the state of the universe. He was looking for a solution that would indicate a static universe. But no matter how hard he tried, he couldn't find one. All of his calculations seemed to show that the universe was either expanding or contracting.

Finally, in order to make his equations work out, Einstein added a fudge factor which he called the *cosmological constant*. The cosmological constant introduced a repulsive force which would cancel out gravitational attraction at large distances. If such a thing existed, it would have to be a very odd thing indeed. Unlike every other force known to physics, it would increase with distance rather than grow weaker. Of course it had been thrown in for no other reason than to make the answers coincide with his ideas. Nevertheless, the introduction of such a quantity seemed impossible to avoid.

Years later Einstein was to remark to physicist George Gamow that the introduction of this cosmological fudge factor was the biggest blunder he had made in his entire life. For if Einstein had not added this factor to his equations, he would have learned that the universe was expanding years be-

Edwin Hubble, the American astronomer who discovered that the universe is expanding. Photograph by Margaret Harwood. *Courtesy Niels Bohr Library, American Institute of Physics.*

fore that fact was discovered by Edwin Hubble in 1929.

Before we discuss Hubble's discovery and its various implications, it will be necessary to talk a little about how astronomers and cosmologists (a cosmologist works with theories about the structure of the universe) obtain the data they use. Unlike the other entities studied by physics, the universe cannot be placed in a laboratory to be observed. One cannot measure it with meter sticks or take its temperature with thermometers or even bombard it with subnuclear particles to see how it reacts.

The only way to obtain information about the universe is to study the light that is emitted by stars and galaxies. Fortunately, light can provide us with more information than one might suspect. First, it is possible to measure its intensity. This often yields quite a bit of information. Next, one can determine its wavelength. Since in most cases thousands of different wavelengths are emitted, a great deal of information can be obtained. If the light emitted by a star or by some other object oscillates, i.e., increases and decreases with time, we have yet more data. And finally, by looking at objects that are very distant from us, we can see the universe as it was long ago. Since the most distant galaxies that astronomers study are something like ten billion light-years away, by looking at them we can see ten billion years into the past. It has taken that long for their light to reach us.

At one time astronomers had to depend entirely on visible light for the information they needed. However, the visible band is only a small part of the electromagnetic spectrum. There are many other kinds of "light." Today all of these can be detected and measured. Astronomers can study ultraviolet and infrared radiation, X rays and radio waves. In the years since World War II, radio astronomy has become especially significant, leading to discoveries that could never have been made with optical telescopes.

In discovering the "Doppler shift," which told him that the universe was expanding, Hubble based his results on observations of visible light alone; ultraviolet, infrared, X-ray and radio astronomy did not exist in 1929. However, when astronomers began to study the other kinds of radiation, they found that all of them exhibited the same shift of wavelength. Virtually every observation that has been made since 1929 has confirmed Hubble's findings.

Like many of the technical terms scientists use, Doppler shift denotes a principle that is really very simple. It is named for Christian Johann Doppler, a nineteenth-century professor of mathematics who noted that when a source of sound or light is moving away from us, wavelengths seem to get longer. It is easy to see why this is so. Picture the source as something that emits a series of wave crests. If it is moving away, each crest has a little farther to travel before it returns to us than its immediate predecessor had. Thus, the wavelength will be longer than it would be for a stationary source. If, on the other hand, the source is moving toward us, the crests are bunched successively closer together, and a shorter wavelength is observed.

The wavelengths associated with red light are longer than those of blue light. Hence, the light from a luminous object moving away from us becomes redder. If, on the other hand, the object is moving toward us, its light gets bluer. Normally, it is impossible to see this effect; the objects we encounter on earth simply do not travel fast enough to cause wavelengths to change by any significant amount. But if one looks at a star that is approaching or receding at a velocity of thousands of miles per second, the shift can be quite noticeable. Since most (but not all) of the stars outside our galaxy are moving away from us, the Doppler shift is often referred to as a *red shift*.

Then a star that is moving away from us will look red—right? Well, not quite. Although the visible light emitted by the star will be shifted toward the red, some of the ultraviolet it emits will be shifted toward the visible band. This compensation means that there will be no change in color noticeable to the naked eye. But when a star is observed with scientific instruments, things will look quite different.

In order to understand how a Doppler shift is observed and how Hubble was able to make his discovery, let us look at an effect that is described by the somewhat forbidding term *Fraunhofer lines*.

Newton believed that all visible wavelengths were present in the light we receive from the sun. During the early nineteenth century, however, Josef von Fraunhofer, a Bavarian physicist, discovered that this was not quite true. Observing the spectrum of the sun, he noticed that certain wavelengths were missing. Today we refer to these missing wavelengths, which are absorbed by the sun's atmosphere, as Fraunhofer lines. In a photograph of a spectrum they appear as dark lines against a background of colored light.

Since Fraunhofer lines are present in the spec-

A portion of the solar spectrum. The dark lines are Fraunhofer lines, wavelengths which are absorbed by the sun's atmosphere. *Courtesy Hale Observatories.*

trum of every star and galaxy, Hubble was able to use them to make very precise measurements of the Doppler shift. Instead of trying to determine whether the light looked "redder" or "bluer," which, as we have seen, would have been impossible, he studied the Fraunhofer lines and noted the amount by which they were shifted. The change in wavelength enabled him to compute the speed of approach or recession.

Hubble observed that although the stars in our own galaxy, the Milky Way, were moving neither toward us nor away from us, other galaxies seemed to be rushing away at velocities up to thousands of miles per second. Almost without exception their light was shifted toward the red, not toward the blue. To be sure, a few, such as the nearby Andromeda galaxy, were approaching; but this could easily be attributed to random motion or to the motion of the solar system as it revolves around the center of the Milky Way. The overall pattern was clear.

At first the idea that the other galaxies should be running away from ours seems a strange one. After Copernicus placed the sun at the center of the solar system, we no longer viewed the earth (or the earth's galaxy) as the center of the universe. However, if the galaxies seem to be moving away from us, it follows that they are also receding from one another. The universe is expanding.

In other words, discounting random motion such as that of Andromeda, every galaxy is moving away from every other. An analogy that is often used is the comparison with a balloon on which a number of dots have been drawn. As the balloon is blown up, each dot moves away from all the others. Another more picturesque analogy compares the universe to a loaf of rising raisin bread. As the dough expands (i.e., rises), the raisins all move away from one another.

If the universe is in a state of expansion, the apparent speed of recession should be greatest for the galaxies that are farthest away. This is just what Hubble observed. He expressed this relationship in mathematical form by saying that the speed of recession of a galaxy was equal to its distance multiplied by a constant. This is known as *Hubble's Law*.

The velocity of recession of the most distant observable galaxies is a substantial fraction of the speed of light. However, the universe is a very big place. In spite of their great speeds, these galaxies are only increasing their distance from us at about the rate of one millionth of one percent every century. The same figure applies also to galaxies that are closer. Their speed isn't so great, but neither is their distance.

Hubble's law is not exact, but the farther one looks out into the universe, the more accurate his calculations become, because as the speed of recession becomes greater, random motions become small by comparison. A nearby galaxy can move toward us, but a more distant object will be carried along by the expansion.

Hubble's Law applies only to galaxies as a whole, not to the stars within a galaxy, which are bound to

one another by gravitational attraction. As the universe expands, each individual galaxy remains the same size; it is the distance between the galaxies that increases. The motion of individual stars is somewhat like that of planets in a solar system. Our sun, in fact, is revolving in an orbit around the center of our galaxy, the Milky Way.

Hubble's Law seems to imply that at some time the universe had a beginning. If the universe is expanding, and if this expansion has always been going on, it must have begun in a very compressed state.

One might think it would be impossible to determine the state of motion of the universe a long time ago. But this is not true. When we observe galaxies that are billions of light-years away, we are looking billions of years into the past. It should be a fairly straightforward matter to project backward and to determine how much time has passed since the expansion began.

When this calculation was first made, a result of 1.8 billion years was obtained. Other data seemed contradictory, however. In particular, radioactive dating of rocks from the earth showed that some of them were 3 billion years old. It appeared that there was something wrong with the expanding universe theory: one can't have 3-billion-year-old rocks in a 1.8-billion-year-old universe!

The central region of the great galaxy in Andromeda, our nearest galactic neighbor. Although the light from most galaxies is red-shifted because they are moving away from us, the light from Andromeda shows a blue shift instead; its random motion is causing it to approach our own Milky Way galaxy. *Courtesy Hale Observatories.*

Fortunately, we can now measure intergalactic distance more accurately than was previously possible and the contradiction has been resolved. We now know that the universe is something like 20 billion years old, not 1.8 billion. For a while, however, it appeared that it might be necessary to find an alternative to, or a modification of, the expanding universe theory.

One such alternative was the so-called *steady state theory* proposed by the astronomers Herman Bondi, Thomas Gold and Fred Hoyle in 1948. Today the steady state theory is all but discredited, as all the evidence seems to indicate that the universe did begin in a state of extreme compression. The steady state theory, however, deserves at least a brief discussion. It was the only important competitor to the now orthodox *big bang theory*, and it does have a certain intuitive appeal; a simple kind of steady state universe is what Einstein was trying to describe when he added the cosmological fudge factor to his equations.

The steady state theory did not deny that galaxies were moving away from one another. Nevertheless, it said that the universe has always been essentially the same, that 10 billion years in the past it had the same appearance it has now, and that it would still look the same 10 billion years in the future. According to the steady state theory, neither the structure nor the appearance of the universe has changed with time.

This may seem to be a contradiction of Hubble's Law. How can the universe remain the same if galaxies are always moving farther away from one another? To claim such a thing sounds like nonsense.

The steady state theory, however, had a way out of the difficulty. It postulated that as the galaxies moved apart, new matter was created between them in the form of extragalactic hydrogen. The name given to this process was *continuous creation*. According to the theory, this new matter eventually coalesced into clouds of gas and then into new galaxies. These new galaxies would be formed just fast enough to fill up the empty spaces left by the expansion of the older ones.

Suppose, for example, two galaxies are a million light-years apart. In time their distance would increase to 2 million light-years. During that same period of time, according to the steady state theory, enough new matter would be created to form a new galaxy between them, so that we would again have galaxies a million light-years apart. No matter how much time went by, the density of galaxies would remain approximately the same.

At first glance the steady state theory doesn't seem to be unreasonable. Under the proper conditions, matter and energy can be transformed into each other. For example, destruction of matter is responsible for the energy released in a nuclear explosion. The reverse process can easily be produced in the laboratory; it is really not very difficult to create matter from the energy of gamma rays.

The steady state theory, however, does not say that new matter is created out of energy—or out of anything else. It assumes that matter is created out of nothing. And since no such process has ever been observed by science, this is quite an assumption.

But it does not disprove the theory. The rate at which it requires that matter be created is so low that it could never be observed. In a space the size of an average physics laboratory, only one new hydrogen atom would have to be created every thousand years to provide the amount of new matter needed. Astronomical distances are so great and the volume of space so enormous that this rate would be sufficient to provide enough material to make the new galaxies that would be needed. On the average, matter is very thinly spread out through the universe. According to the best estimates we have, if all the matter in galaxies were distributed evenly through space, there would be only one atom for every five or ten cubic meters. An alternative way of expressing this is to say that there is one mass the size of the sun for every billion cubic light-years.

For a while it appeared that the steady state theory might be correct. As the years passed, however, and more data accumulated, the steady state theory's major competitor, the big bang theory, began to seem more and more reasonable. Then in 1964 the dramatic discovery was made that it was still possible to see the light from the creation of the universe. Now there seemed to be no alternative but to believe in the big bang. Only by going through unreasonable theoretical contortions could one avoid the conclusion that the universe had originally been in an extremely compressed state and that it had a beginning at a definite moment in time.

To speak of the "creation" or of the "origin of the universe" makes it seem that we are no longer discussing physics but are trespassing into the field of theology. Therefore, a word of explanation is necessary. When scientists talk of the origin of the universe, they mean the beginning of the universe as we know it. They do not mean to imply that nothing existed before that point in time.

Cosmologists can take the expanding universe and extrapolate backward. Eventually they reach a point where the universe is compressed into a very small volume. They find that from here they can go no further, because all their equations cease to work. There is simply no way of knowing what was happening before this time. For lack of a better term, we sometimes refer to this point in time as the "creation." It is possible to make guesses about what might have been happening before the creation, but there is no way to back up those guesses with scientific fact. It may be that we will never be able to do any more than speculate. At the present state of knowledge we can't even be sure that this conceptual barrier will be permanent.

According to the big bang theory, then, the universe "began" with the explosion of a primordial fireball. Perhaps other universes existed before. Perhaps there is no such thing as "before" the explosion. Time itself may have been created in the big bang. We know only that there was an explosion, and that suddenly all of space was filled with light—both visible light and other forms of electromagnetic radiation. The analogy with Genesis 1:3 has, incidentally, often been commented on.

Initially, matter existed, too, but the light played a more significant role. It has been calculated that there were approximately a billion photons for every nuclear particle. This is the same ratio we have today. At the present time, however, matter is the more important, because it is concentrated into galaxies, whereas the photons are thinly distributed through space.

In the primitive fireball, however, it was light that dominated. Matter was only a bystander. During the first 700,000 years, referred to by scientists as the *radiation dominated era* this light was so intense that stable atoms were not allowed to form. As soon as they came together, they were blasted apart.

The billions of years that have passed since then constitute the *matter dominated era*. The primordial light is still present, but it is no longer intense enough to play an important role. It is no longer even visible light. During the eons that have passed since the creation, it has degenerated into radio waves which are given the somewhat unimpressive sounding name, "Cosmic Microwave Radiation Background." Because this can be detected, it is quite accurate to say that we can still see the light from the creation.

Physicist George Gamow, together with Ralph A. Alpher and their collaborator Robert Herman, proposed the big bang theory in the late 1940s. Shortly after their theory was published, Alpher and Herman made some calculations to determine what had happened to the light from the primordial fireball. They concluded that although it would have undergone an enormous red shift, it should still exist in the form of microwaves,* and that the microwave radiation should still be present everywhere. One should be able to see the radiation by looking in any direction in the sky. In other words, it should be present as a background, whence the name Cosmic Microwave Radiation Background.

*Microwaves are radio waves with wavelengths of less than one meter. They were so named because their wavelengths are shorter than those of the VHF band used by radar at the beginning of World War II.

George Gamow, the Russian-born physicist who worked out the "Big Bang" theory of the creation of the universe. *Courtesy Niels Bohr Library, George Gamow Collection.*

For some reason this prediction of Alpher and Herman didn't receive the attention it deserved. When the microwave background was discovered in 1964, the radio astronomers who found it hadn't been looking for it, and they didn't even know what their results meant until they had consulted with some experts in cosmological theory. Communication between theorists and experimentalists in the field of astronomy during this era apparently was not very good.

When radio astronomers Arno A. Penzias and Robert W. Wilson set up their antenna, they intended to make accurate measurements of the intensity of radio waves emitted within our own galaxy. For this purpose they had obtained equipment more sensitive than any that had been used before.

Penzias and Wilson immediately ran into problems. No matter how hard they tried, however much effort they put into checking over their equipment, they were unable to eliminate a persistent background noise or "static" that was spoiling their results. At one point they thought they had traced the cause of the problem to a "white dielectric material" left by pigeons that had been roosting in the antenna. But even after they had cleaned out the bird droppings, the annoying background noise persisted.

To their surprise Penzias and Wilson discovered that the background noise was independent not only of the direction in which the antenna was pointed, but also of the time of day and the season of the year during which the observations were made. Since the cause of the trouble seemed not to lie in their equipment, they began to suspect that they might have stumbled onto a finding of some significance.

The mystery of the microwave background was finally cleared up when Penzias happened to mention the problem in a phone call to astronomer Ber-

Robert Wilson (left) and Arno Penzias. Wilson and Penzias received the 1978 Nobel Prize for their discovery of the Cosmic Microwave Radiation Background. This discovery confirmed the Big Bang theory of the origin of the universe. *Courtesy Bell Laboratories.*

nard Burke. Shortly before, Burke had heard of a talk given by Princeton theorist P. J. E. Peebles in which the latter had mentioned that there ought to be a background of radio noise left over from the big bang, but that it had never been detected.

Penzias got in touch with Peebles and his collaborators, and it was agreed that Penzias and Wilson would publish their results in the *Astrophysical Journal*, and that the Princeton theorists would submit a companion piece giving their theoretical interpretation. Penzias and Wilson gave their paper the cautious title "A Measurement of Excess Antenna Temperature at 4,080 Mc/s." The theorists explained that this "excess antenna temperature" was an observation of light from the big bang. Suddenly the steady state theory began to seem very doubtful; it could not account for a microwave background.

Today there is so much other evidence to support the idea that the universe began with a big bang that it has become the accepted theory. Since we cannot travel billions of years into the past and actually observe what happened, we cannot be absolutely certain that a big bang took place; we must rely on chains of inferences. However, there are really no good theoretical alternatives.

The necessity of a big bang can be avoided if one can reasonably make certain assumptions. For example, one could assume that the strength of gravity varies with time. Perhaps, as the universe ages, gravity grows progressively weaker. It is even possible to speculate that the universe may not be expanding at all. Perhaps the red shift can be attributed to a "tired light" effect which makes light become redder as it ages while traveling across space.

But these are *ad hoc* hypotheses which do not

lead to consistent theories. In particular, none of them can explain the cosmic microwave background. The big bang theory fits the facts as no other does.

It is always possible to invent hosts of explanations for any given phenomenon. Such explanations can be taken seriously, however, only if they also provide explanations for other observed facts. The success of a theory in physics is measured by its ability to tie together a lot of apparently unrelated experimental data. Since the big bang theory does this and the alternatives do not, most scientists agree that the latter should be ruled out as unacceptable.

If we accept the fact that the universe began with a big bang, we may well ask what will happen to it in the future. Will the expansion go on forever? Or will the universe finally slow down enough to reverse itself and enter into a phase of contraction?

We do not know. Gravitational attraction does act as a retarding force that causes the expansion to slow down, but we don't know whether it will ever slow the expansion enough to stop it. The uncertainties are far too great.

If we recall that for our information about the universe we must depend on observations of light emitted from distant galaxies, it is easy to see how these uncertainties arise. The speed of recession of a distant galaxy can be determined, with a reasonable degree of accuracy, from the red shift. Ascertaining its distance, on the other hand, is much more difficult and can only be done by making observations of its apparent brightness. The distance of nearby stars can be determined with great accu-

A faint cluster of galaxies in the constellation Pisces at a distance of well over 4 billion light-years. The light by which this photograph was made had been traveling through space for more than 4 billion years. *Courtesy Hale Observatories.*

racy, but the methods used will not work on galaxies that are millions or billions of light-years away.

If we must depend on measurements of brightness alone, a number of factors can introduce error. If, for example, the galaxy is either brighter or dimmer than the average, our estimate of its distance will be off. Furthermore, we can never see a distant galaxy as it is now; we can only look at light that was emitted long ago and has spent millions or billions of years traveling through space. If the galaxies did not have the same brightness in the past that they have at present, another error will be added to our distance estimates.

We need to know both the velocities of the

galaxies and their distance from us to estimate the rate at which the expansion is slowing. Unfortunately, we simply cannot determine distances with enough accuracy to make this possible.

In astronomy, errors in distance estimates can be quite large. Just such an error in Hubble's original data led to a calculation of the age of the universe that was approximately ten times too small. The figure was revised upward from 1.8 billion years to 20 billion years when the distance measurements were corrected. Yet the present figure could still be in error by as much as 25 or 50 percent.

Another illustration of the uncertainties that plague astronomy is provided by a number that scientists call the *deceleration parameter*. The figure is used to measure the rate at which the expansion of the universe is slowing. If it turns out to be greater than one-half, the universe will eventually begin to contract. If it is less than one-half, the expansion will continue.

The best value for the deceleration parameter is a figure of one. We can't conclude from this, however, that a phase of contraction will eventually be reached. The uncertainty in the determination of the deceleration parameter is approximately 100 percent; the correct value could be anywhere between zero and two. And the determination is not likely to become appreciably more accurate in the near future.

Other experimental data indicate that the expansion will go on forever. Whether this is true depends on estimates of the amount of mass in the universe. One often hears cosmologists speak of *missing mass*; that is, mass which might be present in the universe but which we are not able to observe. Such missing mass could conceivably exist in the form of invisible black holes or interstellar gas.

We don't even know whether the universe is infinite or finite. Einstein's general theory of relativity provides methods for answering such a question, but first we must solve the problem of the slowing of the expansion. The two are bound up with each other. All we can say is that if the expansion goes on forever, the universe is infinite; if the expansion is slowing enough so that the universe will eventually go into a contraction, we can conclude that it is closed and finite.

But suppose the universe does eventually begin to contract. Will it then be possible to say anything about its ultimate fate?

Again the answer is no. We can do no more than guess. If a contraction does set in, it will most likely continue until the universe is so compressed that we will have something very much like the primordial fireball of the big bang. However, we cannot say what is likely to happen next. The universe may be destroyed in this final compression. Or the process could reverse itself, and the universe could be reborn in another big bang.

It is just possible then that we live in an oscillating universe which goes through alternating cycles of expansion and contraction over periods of hundreds of billions of years. But if this is the case, we can only speculate as to whether each cycle would then duplicate the last. It is possible that the stars and galaxies would be recreated in each successive phase. Conditions might be such that life would again be created, and evolution could again produce intelligent life resembling man.

Or something entirely different could take place. A reborn universe might have laws of physics entirely different from those we know. There might be no chance at all that life would ever evolve; there might not even be planets or stars.

In other words, physics tells us nothing. It does not tell us whether there will be future universes; it tells us even less about what these universes might be like. It leaves us free to speculate, on the one hand, that we will all be reborn in a series of endless cycles resembling those of Hindu mythology, or, alternatively, that life is a lucky accident that exists for only a short period of time in a very small number of all the possible universes. It leaves us free to guess that the universe may have been created at a definite point in time and that it will eventually die as it expands forever into the cold reaches of space.

As has been noted several times already, all our knowledge of the universe is based on inferences derived from observations of light. We should be happy that this allows us to know as much as it does. No matter what knowledge we gain, there will always be some point where physics has to leave off; science is able to go only so far. There will always be a boundary beyond which speculation must be left to the philosophers and the theologians. And if some philosophers—the logical positivists, for example—tell us that such speculation is meaningless, we aren't likely to stop it. Asking "meaningless" questions has always been one of humanity's favorite activities.

Nine:
Relativity and Black Holes

When Maxwell discovered that light consisted of electromagnetic vibrations, he proposed that all of space was filled with an invisible substance called the *ether*. The existence of an ether was an ancient idea dating back to the Greeks, and Maxwell and other nineteenth-century physicists thought nothing could be more logical. The existence of oscillations implied that there was some substance to carry the vibrations. Just as air was the medium which supported the vibrations of sound, they said, there had to be an ether to carry those of light.

When theoretical physicists hypothesize the existence of an entity, the next step is always to attempt to detect it experimentally. So in 1887 two physicists, Albert Michelson and Edward Morley, set up an experiment which they were sure would show that an ether existed.

Michelson and Morley reasoned that the earth was moving around the sun at a velocity of about thirty kilometers per second. This should give rise to an "ether wind" flowing past the earth in the opposite direction at the same speed. It should be possible, therefore, to detect differences in the speed of light when it was traveling with the wind and when it was traveling against it. When the light was going "downstream," it should seem to go thirty kilometers per second faster; when it went "upstream," its velocity should appear to be thirty kilometers per second less.

To Michelson's and Morley's astonishment, the experiment detected no ether wind whatsoever. Their surprise was shared by scientists around the world, and before long physics was entering another of its periodic crises. Attempts were made to salvage the ether theory by suggesting that the earth somehow dragged the ether along with it. Other ingenious explanations were devised, but none of them was entirely acceptable.

Albert Einstein. *Courtesy of AIP Niels Bohr Library.*

Then, in 1905, a twenty-six-year-old physicist named Albert Einstein dropped a bombshell. The concept of an ether, he said, was superfluous; the Michelson-Morley experiment had failed because there was no ether wind. Einstein theorized that light always traveled through empty space with the same velocity, independent of the motion of the emitting body. With this simple assumption and the postulate that the laws of physics should be the same in all frames of reference (this Einstein called the *Principle of Relativity*), Einstein was able to develop a theory that startled the world—the *Special Theory of Relativity*.

Einstein's assumption about the constancy of the speed of light was by no means obvious; in fact, it seemed to defy common sense. To find the speed of a bullet fired from a jet plane, for example, we add the bullet's velocity to that of the jet. If the plane is traveling at a speed of 600 miles per hour and if the muzzle velocity of the bullet is 800 miles per hour, the latter will travel through the air at a speed of 1,400 miles per hour. Einstein was saying that this situation does not hold true for light.

But this was only one of many strange ideas inherent in this special theory. It also implied that the velocity of light was independent of the motion of the observer. For example, suppose that a spaceship is traveling toward the sun at a very high speed. If the passengers on the ship measure the velocity of the light coming from the sun, they will find that it passes them at 186,000 miles per second. If they turn the ship around and accelerate until it is speeding in the opposite direction, the light from the sun will still pass them at 186,000 miles per second. Light seems to go no faster when they are rushing to meet it than when they are running away from it. To anyone who has ever watched another car slowly pass him on the highway, this seems very strange.

Special relativity made other predictions that seemed even more bizarre. It said that if an object attained a velocity approaching that of light, that object would appear to flatten out and to become heavier simultaneously. Meanwhile, time as experienced by the object or anyone on it (for instance, a spaceship) would seem to slow down.

We are concerned with relativity primarily as it relates to the subject of light. However, the last-mentioned effect, known as *time dilation* (sometimes as *time dilatation*), is so often used in science fiction and mentioned in discussions of the possibility of interstellar travel that it deserves a few words of elaboration.

Suppose a spaceship traveled away from the earth to another star at a velocity 99 percent of the speed of light. Then the passengers would age only one year for every seven years that passed on earth. And if they reached a velocity that was 99.995 percent that of light, the ratio would be one year to a hundred years. Thus, if we could only make spaceships that went fast enough, a space traveler could go on a voyage that seemed to him to be only a few years in length and would discover on his return to earth that centuries had gone by.

A limerick first published in *Punch* tells us that

> There was a young lady named Bright,
> Who traveled much faster than light.
>   She started one day
>   In the relative way,
> And returned on the previous night.

Unfortunately, special relativity does not permit such delightful paradoxes. It tells us that the speed of light in a vacuum — 186,000 miles per second — is an upper limit. Neither Ms. Bright nor subatomic particles nor anything else can attain or exceed that velocity. The reason for this is simple.

Relativity states that as an object goes faster, it becomes heavier.* The heavier it is, the more energy it takes to accelerate it (just as a Cadillac gulps down more energy than a Volkswagen). A simple calculation shows that the energy required to accelerate any object to the speed of light would be infinite. Hence, until someone finds an infinite source of energy, nothing will be able to go that fast.

Relativity says that no object can attain or exceed the speed of light *traveling through a vacuum*. The word *vacuum* is an important one. Small particles such as electrons or protons *can* sometimes travel faster than light travels in a material substance. When they do, they sometimes emit a visible glow called *Cherenkov radiation*.

When light passes through matter, it slows down. For all practical purposes, its velocity in air is the same as in empty space — 186,000 miles per second. In water, however, the speed of light is only 140,000 miles per second, and in glass it is generally 120,000 to 130,000. Relativity will not allow anything to go as fast as 186,000 miles per second, but speeds of 120,000 to 140,000 can easily be exceeded.

There are other kinds of faster-than-light particles; hypothetical ones called *tachyons* were pro-

---

*To be precise, one would have to say that it seems to become heavier when viewed from a stationary reference system. The ensuing discussion is not affected by this, however.

posed in the mid-1960s by Gerald Feinberg and independently by George Sudarshan and his co-workers. The tachyon is a particle whose speed in a vacuum could never be less than that of light. This is not, however, a contradiction of special relativity. If tachyons existed, they would simply meet the same barrier from the other side; they would not be able to travel at velocities less than 186,000 miles per second. Tachyons would have some very strange properties. For example, as they lost energy, they would go faster. If it were possible to lower their energy to zero, they would then have infinite velocity.

Even if tachyons existed—and they are no more than an interesting possibility—they would not solve any theoretical problems in physics. Experiments designed to detect them have not met with success, but this negative result cannot be considered conclusive. Some scientists suggest that tachyons might exist and yet be immune from discovery; it could be that they cannot interact with our world and we cannot come into contact with theirs. If so, the very idea of the existence of such particles becomes almost metaphysical.

We're fairly safe in assuming, therefore, that the velocity of light is the upper limit of anything we can detect. This assumption has a very important implication. We have noted that an object becomes heavier as it goes faster. Or, in more precise terms, its mass increases. At the same time, the object gains energy (the energy that has been put into it to make it go so fast). By simply writing down mathematical expressions for the mass and energy in terms of an object's velocity, Einstein was able to derive the famous formula

$$E = mc^2$$

where E is the energy, m is mass, and c^2 is the velocity of light multiplied by itself. Since c is 300,000,000 meters per second, c^2 has a value of 90,000,000,000,000,000 (9×10^{16} in the notation used by scientists). It is obvious that mc^2 is a very large number. This accounts for the enormous amount of energy released in explosions of nuclear bombs, and for the energy produced by the sun as well. In both cases energy is created when matter is destroyed.

There is a popular misconception to the effect that the equivalence of matter and energy was initially confirmed in 1945 when the first atomic bomb was detonated at Alamogordo, New Mexico. In reality Einstein's theory received experimental confirmation many years before that, when it was demonstrated that matter could be created directly from the high energy light referred to as *gamma radiation*.

In 1930, when the English theoretical physicist Paul Dirac published a theory combining relativity and quantum mechanics, it had long been believed that matter was converted into energy in the process of radioactive decay. This was only an inference, however. It was known that when a radioactive atom decayed, the products of the decay weighed a little less than the original material. Although it seemed that matter must have been converted into energy, there was no way to observe the process directly. Furthermore, in one type of reaction—beta decay—the correct amount of energy was not obtained, and a new particle, the neutrino, had to be invented to account for the discrepancy. No one had ever suggested that the reverse process could take place, that matter could be made from energy.

It came as a great surprise, therefore, when Dirac asserted that pairs of positively and negatively charged electrons could be created out of gamma rays. At the time, no one had ever seen a positively charged electron; only the ordinary negatively charged variety was known. Furthermore, the only positively charged particle known to physics was the proton. Dirac's theory was therefore greeted with skepticism. There were many physicists who, even though they accepted Einstein's formula, refused to believe that matter could be created out of energy.

Dirac was vindicated, however, in 1932 when Carl Anderson, working in a laboratory at the California Institute of Technology, photographed the track of a strange new particle in a cloud chamber. Measurements of the deflection of the particle in a magnetic field indicated that it must have the same mass as an electron and that it must be positively charged. The new particle was named the *positron*. Dirac's positive electron had been found.

The positron is the electron's *anti-particle*. Today it is known that every particle has an anti-particle. There are anti-protons, anti-neutrons, even such strange beasts as anti sigma hyperons. Particles and anti-particles can be created in pairs whenever there is enough energy. Every time that such "pair creation" takes place, Einstein's $E = mc^2$ receives still another experimental confirmation.

Special relativity is one of the best-established theories of modern physics. Proof of its validity is obtained every day, whenever pair creation is observed and whenever subatomic particles increase in mass as they are accelerated to high velocities in particle accelerators. Yet special relativity has been the object of more skepticism and criticism on the part of nonscientists than any other theory in physics. The days are not long past when one would sometimes hear that someone had dreamed up arguments purporting to show that special relativity just couldn't be true, and when such ridiculous statements as "It is said that only five men in the world can understand the theory of relativity" appeared in the popular press.

Special relativity is unique among the theories of modern physics because for some reason it gained a great deal of notoriety soon after it was propounded in 1905. Ever since it was published, journalists with little understanding of physics have been trying to explain it. As a result, misinformation is still widespread.

It might be well, therefore, to emphasize several points. First, the fact that it is called a theory does not mean that it is not well established. In physics, the word *theory* is frequently applied to an idea that has been well confirmed, whereas the term *hypothesis* is used to describe something that no one is really sure about. There are exceptions to this, of course, as scientists can be as sloppy in their use of language as anyone else. One can state, however, that scientists today no more doubt special relativity than they doubt that the earth is round.

Second, special relativity is not particularly difficult or arcane. Mathematically, it is much less complex than most theory; little more than high school algebra is needed to derive its major results, including $E = mc^2$. The achievement of Einstein lay not in his ability to manipulate complicated mathematical entities (although he could do this), but in his willingness to make that simple but not very obvious assumption that light has a constant velocity. Or, as Einstein himself put it in the paper on relativity published in *Annalen der Physik* in 1905, "Light is always propagated in empty space with a definite velocity *c* which is independent of the state of motion of the emitting body."

Finally, the effects that relativity predicts—the increase of mass with velocity, the slowing down of time, and so on—really aren't especially odd; they are only unfamiliar. If we habitually moved around at velocities approaching that of light, they would seem to be no more than common sense. The difficulty in accepting such ideas lies in the fact that relativistic effects, like photons, are not a part of our everyday world. Once the initial conceptual barrier is passed, there is no problem.

There is another theory of relativity that is very complex: Einstein's theory of gravitation, called the *general theory of relativity*. It is so complicated, in fact, that Einstein had to stop work on it for a while in order to search for the mathematical tools he needed to complete it.

The general theory of relativity, often called *general relativity*, introduces such concepts as four-dimensional *space-time* and *curved space*. Even theoretical physicists cannot visualize the latter; they can only describe it by using a complicated kind of mathematics called tensor analysis.

Einstein published the general theory of relativity in 1916. Three years later a dramatic experimental "proof" of the theory was obtained during a total eclipse of the sun. Although this experimental confirmation received wide publicity in newspapers, it was somewhat less than convincing, as we shall see.

Einstein had predicted that if a ray of light passed close enough to the surface of the sun, the ray would be deflected by the sun's gravitational field. This was nothing new; one could obtain a similar prediction from the Newtonian gravitational theory. General relativity and Newtonian theory, however, gave deflections of different amounts. If the degree of deflection could be measured, the results would show which theory was correct.

Einstein's prediction could not be tested under normal conditions. Light from the stars grazes the surface of the sun all the time, but the sun is so bright that we cannot distinguish it. During an eclipse, however, this bright background is obscured. It should then be easy to test the theory. If the light is bent, the position of the star from which it comes will appear to change. It is only necessary to determine the position of a star before and during an eclipse.

A total eclipse of the sun occurred in 1919, and an expedition of scientists headed by the English astronomer Arthur Eddington went to Africa to observe it. Einstein's theory had predicted a deflection of 1.75 seconds of an arc. Two measurements made in 1919 showed deviations of 1.98 and 1.61, giving an average of 1.80, which was certainly very close. Einstein's theory had been confirmed. Or had it? The tests had been so difficult and the effects measured so small that no one was really sure.

The quantity 1.75 seconds of arc isn't very large. In fact, it is only five ten-thousandths of one degree. It was characteristic of general relativity that the effects it predicted were so small that it was hard to

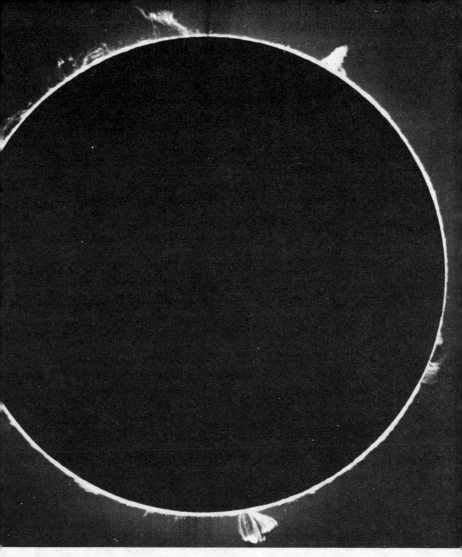

The first experimental confirmation of Einstein's general theory of relativity was attempted by a group of astronomers headed by Sir Arthur Stanley Eddington in 1919. During a total eclipse Eddington and his co-workers observed starlight that grazed the surface of the sun. The experiment could not have been performed at any other time, since under ordinary conditions sunlight would blot out light from the stars in the sun's vicinity. *Courtesy Hale Observatories.*

tell whether or not the various tests confirmed the theory. Physicists admired the general theory as an intellectual achievement; so imposing was its theoretical structure that many of them felt that it simply had to be true. But given the paucity of experimental evidence, it was impossible to be absolutely sure. The trouble was that general relativity gave such minute corrections to Newtonian gravitation; when one is measuring very small quantities, there is always room for error.

For nearly fifty years general relativity was the ugly duckling of theoretical physics. Although it was widely admired, its applications were few and its experimental tests not entirely convincing. As a result, alternate theories of gravitation were still being considered as late as the mid-1960s.

But then in the sixties and seventies there was an enormous upsurge of interest in the theory. One cause was the development of new experimental tests which seemed to vindicate general relativity once and for all. Others were the discovery, beginning in 1962, of mysterious objects in the sky called *quasars*, and the speculation about the possible existence of even stranger entities called *black holes*.

Quasars, or *quasi-stellar objects*, are not yet understood. They have red shifts, which seem to imply that they must be very distant from us. Some of them appear to be billions of light-years away, at the very limit of the seeable universe. To be visible at such great distances, they would have to be shining with the brightness of a hundred galaxies; yet the evidence indicates that they are not much bigger than ordinary stars. This means that they would have to be burning up as much energy in a single year as our sun does in its entire lifetime. We have

Quasars are the oldest and the most distant objects that we can perceive in the universe. For years they have been the subject of much controversy among scientists. The pictured quasar is known only as "3C 273." *Courtesy Hale Observatories.*

no idea where the necessary fuel comes from. Some astronomers have concluded that quasars may be much closer than they seem to be. But that assumption—by no means widely accepted—creates insurmountable difficulties of its own.

To make matters even more puzzling, radio astronomers have discovered two objects associated with a quasar that appear to be receding from each other at nine times the speed of light. Unless we are willing to disregard special relativity, this simply cannot be so. All in all, it seems unlikely that the riddles associated with quasars will be solved in the immediate future.

Since we don't know what quasars are, it would be useless to attempt to apply general relativity to them; we simply wouldn't know where to start. Their discovery, however, had the effect of kindling interest in cosmological speculation in general and in the general theory of relativity in particular. This led in turn to quite a bit of speculation about the hypothetical entities known as black holes.

The idea of a black hole, with a gravitational field so strong that nothing, not even light, can escape it was not a new one. It had first been suggested as far back as 1798 by the French mathematician Pierre Simon de Laplace. In the twentieth century, astrophysicists discovered that, according to Einstein's hypothesis, if a dying star were big enough, it could theoretically become so compressed that a black hole would be formed.

Black holes are just that: black holes in the fabric of space that suck up everything in their paths — light, other kinds of electromagnetic radiation, particles of matter. They are able to do this because of the ways in which they curve space.

Described by the general theory of relativity, the concept of curved space explains why such bodies as planets follow certain orbits. According to general relativity, the earth moves around the sun not because the sun exerts any force on it, but because the sun's large gravitational field curves space within its vicinity. Similarly, it is not the pull of gravity that causes an apple to fall to the earth, but the way in which the earth curves space.

This in no way implies that physics has done away with Newton's law of gravitation. Newton's law is a very close approximation, one that is accurate enough for almost all practical purposes, including the calculation of the trajectories of space probes to Jupiter. But when we begin to deal with very large gravitational fields, such as those in or near black holes, Newton's law no longer works. It is necessary to speak of curved space instead.

Curved space cannot be visualized. However, there is an often-used analogy that makes the concept a little easier to grasp. Imagine a rubber sheet that is stretched out flat like a trampoline. A weight placed on this sheet will cause an indentation. Now place a marble on the sheet near the weight. The marble will begin to roll toward the weight, not because the weight exerts any force, but because "space" (i.e., the rubber sheet) is curved. By giving the marble the right kind of push, we can cause it to circle around the weight in a manner analogous to the way in which a ball spins around the rim of a roulette wheel or the earth orbits the sun.

General relativity tells us that if a dying star has a mass three times greater than that of the sun, it can undergo a collapse which condenses it so greatly that gigantic gravitational fields are created. As the collapse continues, these fields become so strong that nothing, not even light, can escape them. Using the curved space model, we would say that the black hole deforms space so strongly that space curves in upon itself, isolating the condensed star from the rest of the universe.

We can't be absolutely sure that black holes exist, even though general relativity gives us very good reasons for thinking they do. One of the problems is that since by definition a black hole emits no light, it can't be observed directly. Even if we could manage to turn a searchlight on it, that wouldn't help; the black hole would simply suck up all the light and we still would not be able to see it.

Fortunately, there are indirect methods which should enable us to "see" black holes. A large number of solar systems contain double stars. If one of the stars has become a black hole, we should be able to deduce its presence by its effects on its companion. In particular, large amounts of X-ray radiation should be emitted when matter from the visible star falls into the black hole.

At the moment, the most likely black hole candidate is a source called Cygnus X-1 in the constellation Cygnus (the Swan). Cygnus X-1, a powerful emitter of X rays, is believed to be a system of two objects, one of which is a supergiant star. Some astronomers believe its invisible companion is a black hole and that the X rays are produced by the interactions between the two bodies. Others say the X rays could be produced in other ways.

Some astronomers think the universe may contain billions of "mini black holes" which could have been formed shortly after the big bang that marked the beginning of the universe. Some of these would be no bigger than a virus, but would weigh millions of tons. It has been conjectured that the gigantic explosion that rocked Siberia in 1908 could have been caused by a black hole the size of a dust particle. It might have struck the earth, causing the explosion, and then have continued straight through our planet, emerging again on the other side to continue its journey through space. Although many scientists believe that there are good reasons to

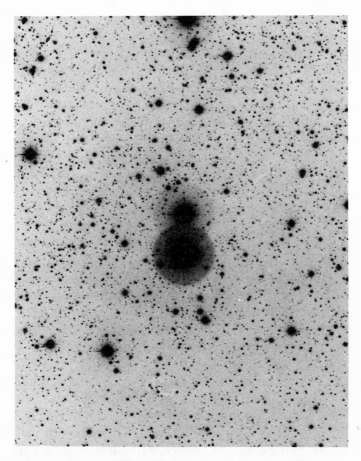

The massive visible star of Cygnus X-1. It is believed to have an invisible companion, a black hole. As the black hole draws material from its visible partner, X rays are produced which can be detected by satellites. *Courtesy Dr. Jerome Kristian and Hale Observatories.*

doubt such an explanation, the possibility cannot be entirely ruled out.

Since a black hole is formed when matter and energy collapse into a very small volume at the center of the hole, it seems logical to ask, "Is this the end of the story, or is there some way that a black hole can re-explode?" This isn't an easy question to answer. Mathematical calculations suggest that if a black hole did explode, some of the particles within it would have to go backward in time! This strange conclusion is tied up with an even more bizarre conjecture: that the matter and energy in the black hole pass through a "wormhole" and emerge as a brilliant *white hole* in a different part of the universe or in another universe entirely. (Mini black holes are an exception to this; there are reasons for thinking they can explode without going through any such contortions.) It has been suggested that supernovas might be white holes produced either by the collapse of black holes in another universe or by the delayed explosion of mini black holes left over from the big bang. It has also been suggested that white holes might provide an explanation for quasars.

Now, one can't accept such ideas without experimental proof. Until some kind of confirmation is forthcoming, the only reasonable course is to keep our reservations about such speculative ideas. On the other hand, we shouldn't reject them just because they sound "crazy." In physics it is the crazy ideas that sometimes have the best chance of being true. The physicist Freeman Dyson made this point in an article called "Innovation in Physics." Dyson recalled that in 1958 physicists Werner Heisenberg and Wolfgang Pauli put forward an unorthodox theory of subnuclear particles. Pauli presented their hypothesis to a group of scientists that included Niels Bohr. In the discussion that followed, some of the scientists were critical of the theory.

Finally Bohr spoke. "We are all agreed," he said,

"that your theory is crazy. The question that divides us is whether it is crazy enough to have a chance of being correct. My own feeling is that it is not crazy enough."

Naturally we can't conclude just from the craziness of the idea that white holes and alternate universes must exist. On the other hand, it might be that the truth is something even more bizarre.

As we pointed out previously, the Newtonian theory of gravitation is good enough to be used under most circumstances. But when gravitational fields are very strong, it becomes necessary to work with general relativity instead. It is possible that the gravity inside black holes is so intense that general relativity will have to be modified. In order to describe accurately the formation of a black hole, we may need a theory that combines general relativity and quantum mechanics. Unfortunately, no such theory exists. Theoretical physicists have no idea even how one would go about merging the two theories; they have no idea what a combined theory would be like or what it might tell us about black holes. If someone does succeed in making the two compatible, however, we can be reasonably sure it will lead to some ideas which seem very "crazy," perhaps even more so than the ones we have been discussing.

Astrophysical arguments suggest that by now as many as one percent of all stars may have evolved into black holes. If this is true, there could be something like a billion black holes in our galaxy alone. In addition, giant black holes with masses hundreds of thousands of times greater than that of our sun might exist in the centers of galaxies. As these gobble up the matter and energy in their surroundings, they will grow even larger. Hence, there is a possibility that our entire galaxy might eventually be converted into an enormous black hole.

If such a thing ever happens, it will not be until long after our sun has become a dead star, billions of years in the future. At that time, in all probability, the human race will have long since ceased to exist, so we can't consider it an especially frightening possibility. Unlike those discussed in Chapter 2, the Lords of Darkness that science has discovered in the universe seem to present no imminent threat.

Bibliography

Chapter 2

Allegro, John. *The Dead Sea Scrolls*, 2nd ed. Harmondsworth, Middlesex, England, and New York: Penguin Books, 1964.

Brandon, S. G. F. *Religion in Ancient History*. New York: Charles Scribner's Sons, 1969.

Campbell, Joseph. *The Masks of God: Occidental Mythology*. New York: Viking, 1964.

————. *The Masks of God: Oriental Mythology*. New York: Viking, 1962.

Cavendish, Richard. *The Powers of Evil*. New York: Putnam, 1975.

Gaster, Theodor H., ed. *The Dead Sea Scriptures*, 3rd ed. New York: Anchor, 1976.

Lucas, George. *Star Wars*. New York: Ballantine, 1976.

Russell, Jeffrey Burton. *The Devil*. Ithaca, N.Y., and London: Cornell University Press, 1977.

Zaehner, R. C. *The Teachings of the Magi*. London: George Allen & Unwin, 1956.

181

Chapter 3

Aristotle. *De Anima*. Oxford: Oxford University Press, 1931.

————. *De Sensu*. Oxford: Oxford University Press, 1908.

Hewish, A., ed. *Seeing Beyond the Visible*. New York: American Elsevier, 1970.

Huygens, Christian. *Treatise on Light*. New York: Dover, 1962.

Mach, Ernst. *The Principles of Physical Optics: An Historical and Philosophical Treatment*. New York: Dover, 1953.

Newton, Sir Isaac. *Opticks*. New York: Dover, 1952.

Plato. *Timaeus*. Harmondsworth, Middlesex, England, and Baltimore, Md.: Penguin Books, 1965.

Rossi, Bruno. *Optics*. Reading, Mass.: Addison-Wesley, 1957.

Saha, A. I. *Theories of Light from Descartes to Newton*. London: Oldbaurm, 1967.

Sharlin, Harold I. *The Convergent Century*. New York: Abelard-Schuman, 1966.

Tolansky, S. *Curiosities of Light Rays and Light Waves*. New York: American Elsevier, 1965.

Chapter 4

Canaday, John. *Mainstreams of Modern Art*. New York: Holt, Rinehart & Winston, 1959.

Courthion, Pierre. *Impressionism*. New York: Abrams, 1977.

Hanson, Lawrence and Elizabeth. *Impressionism: Golden Decade*. New York: Holt, Rinehart & Winston, 1961.

House, John. *Monet*. Oxford: Phaidon, 1977.

Klonsky, Milton. *William Blake: The Seer and His Visions*. New York: Harmony Books, 1977.

Levey, Michael. *From Giotto to Cézanne*. New York and Toronto: Oxford University Press, 1968.

Mathey, François. *The Impressionists*. New York: Praeger, 1961.

Minnaert, M. *The Nature of Light and Color in the Open Air*. New York: Dover, 1954.

Mount, Charles Merrill. *Monet*. New York: Simon & Schuster, 1966.

Muller, Joseph-Émile. *Rembrandt*. New York: Abrams, 1969.

Murray, Peter and Linda. *The Art of the Renaissance*. New York: Praeger, 1963.

Newmeyer, Sarah. *Enjoying Modern Art*. New York: Mentor, 1957.

Nochlin, Linda, ed. *Impressionism and Post Impressionism 1874–1904*. Englewood Cliffs, N.J.: Prentice-Hall, 1966.

Pool, Phoebe. *Impressionism*. New York and Toronto: Oxford University Press, 1967.

Rewald, John. *The History of Impressionism*. New York: Museum of Modern Art, 1961.

Roberts, Keith. *The Impressionists*. Oxford: Phaidon, 1975.

Rosenberg, Jacob. *Rembrandt*. London and New York: Phaidon, 1968.

Selz, Jean. *Turner*. New York: Crown, 1975.

Sewall, John Ives. *A History of Western Art*. New York: Holt, Rinehart & Winston, 1961.

Smart, Alastair. *The Renaissance and Mannerism in Italy*. New York: Harcourt Brace Jovanovich, 1971.

Chapter 5

Gernsheim, Helmut and Alison. *The History of Photography*. New York: McGraw-Hill, 1969.

————. *L. J. M. Daguerre*. New York: Dover, 1968.

Newhall, Beaumont. *The History of Photography*. New York: Museum of Modern Art, 1964.

Scharf, Aaron. *Art and Photography*. Harmondsworth, Middlesex, England, and Baltimore, Md.: Penguin, 1974.

Chapter 6

Cline, Barbara Lovett. *The Questioners*. New York: Crowell, 1965.

De Broglie, Louis. *The Revolution in Physics*. New York: Noonday Press, 1953.

Guillemin, Victor. *The Story of Quantum Mechanics*. New York: Charles Scribner's Sons, 1968.

McKenzie, A. E. E. *The Major Achievements of Science*. New York: Touchstone Books, Simon & Schuster, 1973.

Slater, John C. *Concepts and Development of Quantum Mechanics*. New York: Dover, 1969.

Chapter 7

"The Advance of Laser Weapons." *Science News*, 20 September 1975, p. 191.

Berns, Michael W., and Rounds, Donald E. "Cell Surgery by Laser." *Scientific American*, February 1970, pp. 99–110.

Buchsbaum, Simon J. "Lightwave Communications: An Overview." *Physics Today*, May 1976, pp. 23–25.

Bylinsky, Gene. "Laser Alchemy Is Just Around the Corner." *Fortune*, September 1977, pp. 186–90.

"Cutting Cloth by Laser." *Time*, 22 March 1971, p. 71.

Douglas, John H. "High-Energy Laser Weapons." *Science News*, 3 July 1976, pp. 11–13.

"Have Laser, Will Travel." *Newsweek*, 28 March 1977, p. 87.

"Laser Scalpel." *Time*, 6 August 1973, p. 68.

"A Laser System That Spots Product Defects." *Business Week*, 2 June 1973, p. 80.

"Lasers Come of Age, from Hospitals to Battlefields." *U.S. News & World Report*, 17 February 1975, pp. 68–71.

"Lasers Get a Grip on Fusion Power." *Business Week*, 20 May 1972, pp. 60–62.

"Lasers Measure Moon Distances to 6 Inches." *Science News*, 10 November 1973, p. 292.

Mims, Forrest. "From the Laser's Eye." *Science Digest*, September 1973, pp. 39–44.

"A New Figure for the Cosmic Speed Limit." *Science News*, 2 December 1972, p. 356.

"Next U.S. Superweapon—The Pentagon's 'Light Ray.'" *U.S. News & World Report*, 18 October 1971, pp. 85–87.

"Physics with Lasers: High Resolution Coming of Age." *Science*, February 1972, pp. 739–40.

"Search for Laser Signals from Elsewhere." *Science News*, 17 May 1975, p. 318.

"Space Vehicles Could Be Propelled by Remote Lasers." *Physics Today*, August 1977, pp. 17–20.

"Take a Dead Man's Eyes." *Newsweek*, 1 February 1971, p. 42.

"Toward Laser Weapons in Space." *Science News*, 5 March 1977, p. 158.

"What Lasers Can Do for the Phone System." *Business Week*, 20 May 1972, p. 62.

Yaffee, Michael L. "Lasers Investigated for Space Propulsion." *Aviation Week & Space Technology*, 21 April 1976, pp. 47–54.

Chapter 8

Berry, Michael. *Principles of Cosmology and Gravitation*. Cambridge: Cambridge University Press, 1976.

Cosmology + 1. Readings from *Scientific American*. San Francisco: W. H. Freeman, 1977.

Landsberg, Peter T., and Evans, David A. *Mathematical Cosmology*. Oxford: Clarendon Press, 1977.

Motz, Lloyd. *The Universe*. New York: Charles Scribner's Sons, 1975.

Rowan-Robinson, Michael. *Cosmology*. Oxford: Clarendon Press, 1977.

Weinberg, Steven. *The First Three Minutes*. New York: Basic Books, 1977.

Chapter 9

Berry, Michael. *Principles of Cosmology and Gravitation*. Cambridge: Cambridge University Press, 1976.

Einstein, Albert, and Infeld, Leopold. *The Evolution of Physics*. New York: Simon & Schuster, 1966.

Feinberg, Gerald. *What Is the World Made Of?* New York: Anchor, 1978.

Gardner, Martin. *The Relativity Explosion*. New York: Vintage, 1976.

Golden, Frederic. *Quasars, Pulsars and Black Holes*. New York: Charles Scribner's Sons, 1976.

Gribbin, John. *White Holes*. New York: Delta, 1977.

Shipman, Harry L. *Black Holes, Quasars and the Universe*. Boston: Houghton Mifflin, 1976.

Taylor, John G. *Black Holes*. New York: Random House, 1973.

Chronology

Index

191